CCNA™ JumpStart™

Second Edition

Patrick Ciccarelli

Christina Faulkner

San Francisco ◆ London

Associate Publisher: Neil Edde
Acquisitions Editor: Elizabeth Hurley Peterson
Developmental Editor: Colleen Wheeler Strand
Production Editor: Erica Yee
Copyeditor: Nancy Sixsmith
Technical Editor: David Groth
Design Illustrations: Chris Gillespie
Graphic Illustrator: Jerry Williams
Electronic Publishing Specialist: Kate Kaminski, Happenstance Type-O-Rama
Proofreader: Emily Hsuan, Laurie O'Connell, Dave Nash, Nancy Riddiough, Sarah Tannehill
Indexer: Ann Rogers
Book Designer: Maureen Forys, Happenstance Type-O-Rama
Cover Designer: Archer Design
Cover Illustrator: Archer Design

Copyright ©2003 SYBEX Inc., 1151 Marina Village Parkway, Alameda, CA 94501. World rights reserved. No part of this publication may be stored in a retrieval system, transmitted, or reproduced in any way, including but not limited to photocopy, photograph, magnetic or other record, without the prior agreement and written permission of the publisher.

First edition © 1999 SYBEX Inc.

JumpStart is a trademark of SYBEX Inc.

Screen reproductions produced with Collage Complete.
Collage Complete is a trademark of Inner Media Inc.

TRADEMARKS: SYBEX has attempted throughout this book to distinguish proprietary trademarks from descriptive terms by following the capitalization style used by the manufacturer.

The author and the publisher make no representation or warranties of any kind with regard to the completeness or accuracy of the contents herein and accept no liability of any kind, including but not limited to performance, merchantability, fitness for any particular purpose, or any losses or damages of any kind caused or alleged to be caused directly or indirectly from this book.

This publication is neither affiliated with nor endorsed by Cisco Systems, Inc. "Cisco", "CCNA", "CCDA", "CCIE", "Cisco Certified Network Associate", "Cisco Certified Design Associate", "Cisco Certified Internetwork Expert", and "Cisco Networking Academy" are trademarks owned by Cisco Systems, Inc.

Library of Congress Card Number: 2002112918
ISBN: 0-7821-4174-9

Manufactured in the United States of America

10 9 8 7 6 5 4 3 2 1

For Cecilia, Rebecca, and Olivia. You fill my world with joy and love, and without you my life would be incomplete. Thank you for making all my dreams come true. Te amo.
—CLF

To my wife Julia, thank you for supporting me during the production of this second edition.
—PJC

Acknowledgments

Creating a second edition of a book is a challenge. At one level, a second edition is an opportunity to update the book with new relevant material. Fortunately, the development and editing team believe strongly that this is also the opportunity to improve the quality of the book. Although this makes all of our jobs more difficult, we hope that it will translate into a more meaningful experience for you.

Special thanks go to Elizabeth Hurley Peterson and Neil Edde for identifying the need to have this book updated in a second edition and getting it approved.

We would also like to thank the people at Sybex who made this book look the way it does: Maureen Forys, the book's designer and compositor, and Jerry Williams, our illustrator, who gets special praise for turning scribbled drawings into the professional artwork in this book. Our technical editor, David Groth did an exceptional job at scanning for technical inconsistencies and errors ensuring this second edition would be even better than the first. Colleen Strand, the developmental editor, was flexible, thorough, and did an excellent job at keeping the project on track. Thanks to Erica Yee and Nancy Sixsmith for making the finishing touches in preparation for final print.

Patrick would also like to thank a few people who made this book possible:

> To my brother Joe for alleviating other pressures so I could focus on the second edition. Big thanks to my good friend Susan Kim for pulling me away from the work and taking me surfing one weekend. Tom Sparks assisted me with some great technical feedback. My thanks again to my friends Laura Cheu, Susan Basalla and Dr. Gene Michaels for their support on the first edition of the book.

Contents

	Introduction	xii
Chapter 1 ◆	**What Is Networking?**	**1**
	Networking Basics	2
	Types of Networks	3
	Local Area Networks	4
	Metropolitan Area Networks	5
	Wide Area Networks	6
	Laying the Foundation: The Public Telephone System	8
	The Origin of Networks	9
	Network Communication	9
	ARPAnet	10
	The Internet	11
	The Future of the Internet	11
Chapter 2 ◆	**Communication Models**	**14**
	What Is the OSI Model?	16
	How Data Flows through the OSI Model	18
	Understanding Each OSI Layer	19
	Layer 7: The Application Layer	20
	Layer 6: The Presentation Layer	22
	Layer 5: The Session Layer	24
	Layer 4: The Transport Layer	26
	Layer 3: The Network Layer	28
	Layer 2: The Data Link Layer	30
	Layer 1: The Physical Layer	32
	The OSI Model versus the TCP/IP Model	34
	Application or Process Layer	34
	Host-to-Host or Transport Layer	35
	Internet Layer	35
	Network Interface Layer	35
	Physical Layer	36
Chapter 3 ◆	**Network Architectures**	**40**
	Understanding Network Architectures	42
	Peer-to-Peer Networks	43
	Selecting Peer-to-Peer Networks	44
	Security on Peer-to-Peer Networks	45

Implementing a Peer-to-Peer Network	46
Configuring Windows Peer-to-Peer Networking	48
Configuring Macintosh Peer-to-Peer Networking	55
Clients and Servers	59
Common Server Types	60
Terminals and Hosts	61
Selecting a Client-Server Network	62
Security on Client-Server Networks	62
Implementing Client-Server Models	63
Hybrid Networks	67

Chapter 4 ◆ Network Services and Software — 70

Types of Network Services	72
Choosing a Server	74
What to Look for in an NOS	74
Server Hardware	76
Types of Network Operating Systems	77
Windows 2000 Server	77
UNIX	79
Linux	81
Novell NetWare	83
Mac OS X Server	85
Client Network Configuration	87
Configuring Windows XP Clients	87
Configuring Internet Addresses	89

Chapter 5 ◆ Topologies — 96

Physical versus Logical Topologies	98
Types of Physical Topologies	99
Bus Topology	100
Ring Topology	101
Star Topology	104
Mesh Topology	106
Hybrid Topologies	108
Types of Logical Topologies	109
Logical Bus	109
Logical Ring	110
Wireless LAN Topologies	112

Contents

Chapter 6 ◆ Electricity — **116**
- Types of Electrical Currents — 118
  - Alternating Current — 118
  - Direct Current — 121
- Power Problems and Solutions — 123
  - Static Electricity — 123
  - Excess Power — 123
  - Power Loss — 126
- Data Centers — 128

Chapter 7 ◆ Signaling — **132**
- What Is a Signal? — 134
- Measuring Signals — 136
- Signals and Computers — 137
  - Analog Signals — 137
  - Digital Signals — 138
  - Analog versus Digital — 141
  - An Analog and Digital World — 142
- Understanding Transmission — 144
  - Transmission Types — 144
  - Transmission Modes — 145

Chapter 8 ◆ Network Media — **150**
- Network Media and Connectors — 152
- Copper Media — 153
  - Coaxial Cable — 153
  - Types of Coaxial Cable — 154
  - Shielded Twisted-Pair Cable — 157
  - Unshielded Twisted Pair — 160
- Fiber-Optic Media — 164
  - Fiber-Optic Cable — 164
- Wireless Networking Options — 169
  - Radio Transmissions — 170
  - Microwave Transmissions — 172
  - Infrared Transmissions — 174
- Comparing Network Media — 175

Chapter 9 ♦ Devices — 180

- Extending the Network — 182
- Network Segments — 183
  - Segmenting the LAN — 184
- Network Interface Cards — 185
  - Source and Destination — 186
- Repeaters — 188
- Hubs — 189
- Access Points — 191
- Bridges — 192
- Wireless Bridges — 193
- Switches — 194
  - Virtual LANs — 194
- Brouters — 196
- Routers — 197
  - How Routers Select the Best Path — 197
- Gateways — 199
- Comparing Networking Devices — 200

Chapter 10 ♦ Standards — 204

- What Are Standards? — 206
  - How Does an Idea Become a Standard? — 206
- Major Standards Organizations — 208
  - ISO: International Organization for Standardization — 209
  - IEEE: Institute of Electrical and Electronics Engineers — 210
  - EIA/TIA: Electronics Industry Alliance and Telecommunications Industry Association — 210
  - ANSI: American National Standards Institute — 211
  - ITU: International Telecommunications Union — 212
- IEEE 802: Standards for Local and Metropolitan Area Networks — 213
  - 802.1: LAN and MAN Bridging and Management — 213
  - 802.2: Logical Link Control — 213
  - 802.3: CSMA/CD Access Method — 214
  - 802.4: Token Passing Bus Access Method — 216
  - 802.5: Token Ring Access Method — 216

Contents

802.6: DQDB Access Method	216
802.7: Broadband Local Area Networks	216
802.8: Fiber-Optic Local and Metropolitan Area Networks	217
802.9: Integrated Services	217
802.10: LAN/MAN Security	217
802.11: Wireless LANs	217
802.12: High-Speed LANs	218
802.14: Cable TV Access Method	218
802.15: Wireless Personal Area Network	218
EIA/TIA Structured Cabling Standards	220
568-A: Commercial Building Telecommunications Wiring Standard	220
568-A/UTP	220
569: Commercial Building Standard for Telecommunications Pathways and Spaces	221
606: Administration Standard for the Telecommunications Infrastructure of Commercial Buildings	221
607: Commercial Building Grounding and Bonding Requirements for Telecommunications	222
WAN Connection Standards	223
CCITT/ITU-T WAN Standards	223
ANSI WAN Standards	226

Chapter 11 ◆ Network Protocols — 232

Why Protocols Are Important	234
Understanding Protocol Suites	236
TCP/IP Suite	237
Features of TCP/IP	238
Protocols of the TCP/IP Stack	238
IPX/SPX Protocol Suite	246
Features of IPX/SPX	246
Protocols of the IPX/SPX Stack	247
AppleTalk Protocol Suite	249
Features of AppleTalk	249
Protocols of the AppleTalk Stack	250

Chapter 12 ◆ LAN Design — 256

Preparing for LAN Design — 258
Needs Assessment — 260
    Equipment Inventory — 260
    Facility Assessment and Documentation — 264
    Assessing User Needs — 266
Creating the LAN Design — 267
    Architecture Selection — 267
    Topology Selection — 269
    Device Selection — 273
    Media Selection and Installation — 275
    Testing and Certifying Cable — 276
    Connecting the LAN Devices — 277
    Connecting to a WAN — 278

Chapter 13 ◆ Network Management — 280

Why Network Management? — 282
First Steps in Network Management — 285
    Backing Up Data — 286
    Uninterrupted Power — 290
    Redundancy — 291
Performance Monitoring — 293
    Baselining: Setting the Starting Point — 293
    Analyzing Network Performance — 294
    Monitoring Server Performance — 296
    Documenting Performance — 296
Network Management Systems — 299
    The Management Model for TCP/IP Networks — 299
    SNMP in Action — 300
    Implementing Network Management Systems — 303
Troubleshooting — 305
    Documentation for Troubleshooting — 305
    A Layered Approach to Network Troubleshooting — 306

Chapter 14 ◆ WANs and Internet Access — 312

Wide Area Networks — 314
Connecting to the Internet — 316
    Connecting to Your ISP — 317

Contents

Planning Internet Access	318
Telecommunications Services	319
WAN Technologies	320
Avoiding the Potholes	326
The Rollout	326
Appendix A ◆ Answers to Review Questions	**331**
Chapter 1	332
Chapter 2	333
Chapter 3	334
Chapter 4	336
Chapter 5	337
Chapter 6	339
Chapter 7	340
Chapter 8	340
Chapter 9	342
Chapter 10	343
Chapter 11	344
Chapter 12	345
Chapter 13	346
Chapter 14	347
Appendix B ◆ Acronyms and Abbreviations	**349**
Appendix C ◆ Glossary	**365**
Index	*384*

Introduction

As a prospective Cisco Certified Network Associate (CCNA) or Cisco Certified Design Associate (CCDA), you face a challenging job market. The outlook over the next few years is that hiring will continue to be slow as the economy picks up again. That means you need as many advantages over the competition as possible. Obtaining certification is one way to distinguish you from the rest of the crowd. Of course, unlike the late 1990s, certification alone won't guarantee a job. You will need to complement your certification with real-world experience; demonstrate a willingness to work on a team, and prove that you have the desire to further your skills by going after more advanced certifications.

Is there still a world of opportunity out there for an emerging Information Technology professional? Of course there is. Just because the job market is tight, it doesn't mean companies don't still face difficulties with regard to networking. New technologies emerge and companies need assistance in evaluating the effectiveness of the technology on their business. As a CCNA, you will be able to assist these companies in evaluating their needs, determining the best technology to meet those needs, and implementing their network.

A CCNA-certified individual is considered to have the skills and knowledge necessary for positions such as a Help Desk Engineer or a Field Technician. Your employer would expect you to know how to do low-level network troubleshooting, install and configure Cisco devices in LANs and WANs, and provide some support for network performance and security. As a CCNA candidate, you will be expected to meet 63 exam objectives. Each objective is matched to a skill or concept that you not only need to understand, but will have to apply in a real-world scenario.

If you are serious about networking and you are excited by the challenges that networking presents, then you will be successful in achieving the first level of Cisco certification. This book takes you one step closer to reaching that goal.

Who Should Read This Book?

This book is designed to teach the fundamentals of networking and internetworking to people who are fairly new to networks. In particular, it is geared toward people who are interested in achieving CCNA or CCDA certification, but who don't necessarily have much network experience.

This book is intended to be used either for self study, within an academic setting such as a networking class, or as a resource if you are currently responsible for

Introduction

a small office network. If you are studying for CCNA or CCDA certification on your own, *CCNA JumpStart, Second Edition* offers a thorough introduction to the topic, or it can be used as a companion to a more advanced study guide. This book is also a great text or companion text for students in high schools, community colleges, and other educational institutions.

If you are not sure if you have the skills to begin on your journey toward the CCNA, take a moment to review some prerequisites and complete the self-assessment that follows.

Suggested prerequisites:

- Proficient in using a computer.
- Experienced at utilizing the World Wide Web and email.
- Able to access a network at home, school, or work.

Self-assessment:

- Do you enjoy analyzing a problem and finding the solution?
- Do others look to you for answers to their computer problems?
- Are you able to focus on the details of a subject or problem?
- Are you curious about how networks work?

If you answered "yes" to most of these questions, then you have what it takes to start your education. As you can see, it doesn't take much. Throughout this book, we will take the approach that you are new to the field. That way, you will be able to follow along and learn the necessary terminology and concepts. By the end of this book, you will feel confident in continuing your studies towards the CCNA or CCDA.

What This Book Covers

Becoming certified as a CCNA is a challenging endeavor. The CCNA and CCDA exams have gone through several revisions, so you can be sure that the exams are accurate and thoroughly assess your knowledge. Unlike other certifications, the CCNA assumes a great deal of prior knowledge about how networks work and how they are used in the real world. This book covers a wide range of topics intended to give you an overview of many of the concepts and technologies required for the CCNA and CCDA exams. If you are intending to expand your credibility with any of the advanced Cisco certifications, you'll find that this book provides you a core foundation critical to your growth.

Introduction

To help you past the learning curve, topics are contained in discrete chunks, with lots of illustrations, photos, and on-screen examples to bring concepts to life. Throughout the book, instructive explanations and analogies disentangle complex networking topics, and practical examples show you how technologies are used in the real world. Here's how the book breaks down:

Chapters 1–2 The preliminary chapters provide an overview of networks and introduce the communication models for data communication.

Chapters 3–5 You'll get your hands dirty in these chapters as you explore the different types of networks, learn about network services and software, and discover how networks are physically laid out.

Chapters 6–7 Now that you have some understanding of what makes up a network, you'll get down under the hood of networks to explore the basics of electricity and signaling.

Chapters 8–11 If you are going to build LANs, you need to know about network media, network devices, and the standards and protocols that determine how they all communicate.

Chapters 12–14 The final chapters of the book are a "how to" of LAN design, network management, and wide area networking. You'll find practical solutions to building and managing networks that you can use immediately.

Making the Most of This Book

At the beginning of a chapter of *CCNA JumpStart, Second Edition,* you'll find a list of topics that you can expect to be taught within that chapter. To help you soak up new material easily, we've highlighted new terms and defined them in the margins of the pages. And to give you some hands-on experience, we've provided "Test It Out" sections that let you practice what you've just learned. In addition, several special elements highlight important information:

NOTE
Notes provide extra information and references to related information.

TIP
Tips are insights that help you perform tasks more easily and effectively.

Introduction

WARNING

Warnings let you know about things you should do—or shouldn't do—as you go about the job of being a network administrator.

At the end of each chapter, you can test your knowledge of the topics covered by answering the chapter's Review Questions. (You'll find the answers to the Review Questions in Appendix A.)

We've also provided some special material for your reference. If you're wondering about a networking acronym or abbreviation, turn to Appendix B, where you'll find a comprehensive list of acronyms and abbreviations and what they stand for. Appendix C is a glossary of all the terms introduced in the book.

Cisco Certification

Cisco offers a wide range of certifications that complement a growing networking field. Unlike the early years of networking, when it was believed that one person could have all of the knowledge and skills necessary to support all network technologies, today the industry is more mature. The technologies are more sophisticated, and our expectations for what these technologies can provide to businesses have increased. Cisco has responded to this change by creating several tiers of certification. Cisco organizes certification tiers into three levels of expertise: Associate, Professional, and Expert. As you move through these levels, you can follow one of three "tracks" up the ladder (or, in Cisco's case, up the pyramid): Network Installation and Support, Network Engineering and Design, or Communication and Services.

	Network Installation and Support	Network Engineering and Design	Communication and Services
Associate	CCNA	CCDA	
Professional	CCNP	CCDP	CCIP
Expert	CCIE Routing and Switching		CCIE Communication and Services

Along with the track categories, Cisco also offers several specializations, which are supplementary certifications that focus on a particular area of technology.

The Cisco Qualified Specialist topics include Internet solutions, cable technologies, and security. A specialization gives the certified individual a technology focus to better align their skills with the needs of clients. Specializations include the following:

- Voice access
- Network management
- Security
- LAN ATM
- SNA/IP integration
- SNA/IP network management

NOTE

For more information on Cisco certification, go to http://www.cisco.com and click on Training/Certifications.

Planning Certification

If you are serious about your career in networking, then you should have a clear plan for achieving your goals. Many experienced professionals have made the mistake of acquiring various certifications with no direct link to their professional or personal reality. Working on a certification, much like working toward an academic degree, takes up time and money. Be sure that you have reviewed your goals and aligned your certification study to those goals. Your plan will keep you motivated and on track to see yourself through the whole process.

Here are some suggestions for planning your certification study:

- Identify your professional goals. These may include desired income and job title.
- Review the Cisco career certifications, and select the certifications that best meet your professional goals.
- Determine your training requirements. Select between one or a mixture of self-study books, self-study online, and instructor-led training.
- Complete your training, and perform significant review in the way of lab practice on Cisco equipment. Online labs are available from Cisco and other vendors.
- Take and pass the necessary certification exams.

Introduction

How to Prepare for the Exam

After you have completed this book, there are several things you can do to prepare for the CCNA or CCDA certification. The first is to get yourself a copy of the *CCNA: Cisco Certified Network Associate Study Guide, Third Edition* written by Todd Lammle (Sybex, 2002). You'll find that you will smoothly transition from *CCNA JumpStart, Second Edition* to the *CCNA: Cisco Certified Network Associate Study Guide, Third Edition* without missing a step.

Preparation for the CCNA certification is more about your skills than it is about passing an exam. When you market yourself as a CCNA, you are representing yourself as someone who is capable of designing, installing, and supporting networks. If you are going to be successful in the networking field, you will need to develop your skills on Cisco equipment.

There are several ways in which you can prepare for the CCNA certification:

- Individual study: This can include books like this one and the *CCNA: Cisco Certified Network Associate Study Guide, Third Edition* from Sybex, online documentation, and hands-on exercises.

- Self-paced training: Interactive CD-ROMs and web-based courses are available to help you prepare for the CCNA exam. Again, you still need hands-on experience configuring routers, but this can come in the form of simulations.

- Instructor-led training: This is the more traditional approach to studying, and is well worth it. Although there aren't CCNA-specific courses, with Cisco's commercial training partners, you need to take only the Introduction to Cisco Router Configuration (ICRC) or Cisco Routing and LAN Switching (CRLS) courses to prepare for the CCNA.

- Cisco Network Academy: In 1997, Cisco began a program teaching students in high schools and community colleges about networks, with the idea that, once trained, these same students could help schools maintain their networks. The program is now available throughout the U.S. and in other countries. You can find out more details at http://www.cisco.com/edu.

Certification Resources

In 1999, when the CCNA certification was launched, there were few resources available online. The websites that were available were geared toward the CCIE certification, and were therefore too high-level. Now, you can take advantage of many free and subscription-based services for all levels of certifications. In addition, technical resources for implementing Cisco equipment are readily available. These additional resources can be invaluable during your studies and on the job.

```
www.ccieprep.com
www.ccnaprep.com
http://studyguides.cramsession.com/cramsession/cisco/ccna/
www.certificationzone.com/shortcuts/CCNA/
www.sybex.com/certification
```

Where to Take the Exam

Cisco exams are given at authorized Sylvan/Prometric testing centers and VUE testing centers. To find the Sylvan/Prometric testing center nearest you, call (800) 204-EXAM or go to its website at www.2test.com. To find the nearest VUE testing center, go to its website at www.vue.com, or call (952) 681-3000. Before you register, make sure that you know the testing center nearest you. Once you find a location near you, determine the day and time you can take the exam, and choose an alternate time in case your first choice is unavailable. (The testing centers are usually open from 9 a.m. to 5 p.m.) You can either register by phone or via a secure site with a credit card. The cost for the CCNA exam is $100 so if you have difficulty the first time around you don't have to worry as the cost for a second exam won't break the bank. If you do find that you have to retake the exam, you can do so immediately.

Another important note about testing is knowing what it is like to take an exam at a testing center. If this is your first time taking a certification exam, you may want to pay a visit to the testing site. There can be a big difference between taking an exam at a testing center located in a quiet location off the highway and one that is located on a busy street in San Francisco. It may seem like a burden to visit the site in advance, but it is a minor issue compared to the frustration of taking your exam in an uncomfortable setting. In addition to the location of the testing center, take note of the layout of the center, the ventilation system, and any disruptive office or street noise. When you are deciding where to take the exam you'll want to know that the only obstacle is the exam itself and nothing more.

Chapter 1

What Is Networking?

A network is a system that allows communication to occur between two people or machines. In the world of computer networking, the rules for communication must be well defined. Communicating computers need to know the rules, so—like two people speaking the same language—they can communicate without delay. If the computers don't understand each other, nothing is accomplished; there is no Internet access, sharing of files, or printing; and all work stops. As a future CCNA, it will be part of your responsibility to help prevent this from happening.

In this opening chapter, you'll learn some of the fundamental terms and concepts behind computer networks. Topics include:

  Networking basics

  Types of networks

  The origins of the first networks

  The Internet

Chapter 1

Networking Basics

Networks are used to make work and communication more efficient. A network may connect together computers, printers, CD-ROM drives, scanners, and other equipment. The advantage of having computers and other machines connected together is that people can then pass information back and forth much more quickly. Before computer networks, people had to use cumbersome diskettes to share information, and before that, paper. Another advantage of using networks is that they allow people to share resources. Printers, hard disks, and applications can be shared, greatly reducing the costs of providing these resources to each person in a company.

A computer network is built around the idea that there are senders and receivers. The sender, or *source*, is a computer that wants to send information to another computer. The receiver is the computer that the information is sent to, also known as the *destination computer*. Often, computers are not the only machines communicating on a network. Other machines—such as printers with network capabilities—can also act as senders and receivers. A printer, computer, or any machine that is capable of communicating on the network is referred to as a *device* or *node*.

When devices are participating in communication on a network, they need some way to pass information among themselves. In most networks, cables are used to interconnect devices. Devices may be strung together like Christmas lights, with the cable going from device to device. In another layout, cables connect each device to a central location, like the spokes of a bicycle wheel. The cable usually used in networking is made of copper wires similar to the wires in telephone cable, but of a much higher quality. In addition to copper-wire cables, there are other types of media that can be used in networks, including cables made of glass and plastic. Most recently, network communication has been accomplished through the air using radio and microwave transmissions.

When two or more networks are connected together and able to communicate, it is called an **internetwork**. Internetworking is the capability of different networks to communicate using special hardware and software. Internetworking devices make it possible for two networks to communicate, even if they use different **protocols**.

---

internetwork
Two or more connected networks of similar or different communication types.

protocol
A set of rules used to define communication between two devices.

What Is Networking?

Types of Networks

There are three main categories of networks:

- A local area network (LAN) is a small network of computers and printers in a single building or floor.
- A metropolitan area network (MAN) is a high-speed internetwork of LANs across a metropolitan area.
- A wide area network (WAN) connects LANs using the public switched telephone network.

Local, metropolitan, and wide area networks are quite different from one another. In addition to covering geographic areas of different sizes, the network types have varying installation and support costs associated with them.

Devices used within LANs can be relatively inexpensive and easy to maintain. In many cases, a single person can be responsible for all LAN-related issues. Often in very small offices, one person may take on the responsibilities of network support as an adjunct duty to their regular work. Other small- to medium-size offices hire consultants or a dedicated staff person to provide technical support they cannot provide themselves.

Larger networks, such as metropolitan and wide area networks, require more sophisticated networking equipment and support. The investment in a MAN or WAN is not only based on installation and equipment costs, but also on the costs of long-term support and on-site administration to keep the network running properly. Most larger networks require at least one full-time on-site administrator to maintain the network.

Today, because of the ease of access to the Internet, companies can connect to remote or distant locations without spending lots of money. A person working within a small local network with Internet access can share documents and files with people all over the planet, and even access servers at distant locations. The global reach of the Internet allows this kind of connectivity without the high cost of installation and support associated with private wide area networks.

NOTE

One additional type of network that is found today is the Campus Area Network, or CAN. This specialized network provides services through high-speed fiber-optic connections between buildings containing one or more LANs.

Chapter 1

Local Area Networks

The **local area network (LAN)** plays an important part in the everyday functioning of schools, businesses, and government. LANs save people time, lower equipment costs by centralizing printers and other resources, and allow sensitive information to remain in a secure location. Recently, LANs have been used as tools to improve collaboration between employees and for job training using audio and video.

A local area network is used to connect computers and other network devices together so that the devices can communicate with each other to share resources. Devices on a LAN are connected together using inexpensive cable. Due to limitations in distance, performance, and manageability, the LAN is usually confined to a single office or floor of a building.

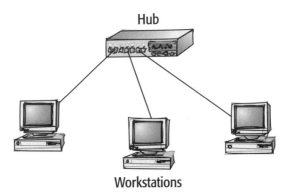

In the preceding illustration, several computers are connected via a cable to a central device called a **hub**, or **switch**. Hubs and switches are common devices found on a network. The lines from the computers to the hub are the cables that allow data transmissions to pass from one computer to the others.

Today, many new local area networks are being installed using wireless technologies. Wireless LANs allow users to connect to network resources without the installation of cabling or wiring. They use wireless devices such as **access points**, or APs, to transmit and receive data.

Depending on the size of the company and the building, there may be one or more LANs. A company that is located in a multistory building with hundreds of employees may have a LAN on each floor. Between each floor, a **bridge** or a **router** is used to interconnect the LANs. Inside the LAN's computers, printers, and other network-capable devices are **network interface cards** (**NICs**) that allow the devices to communicate at any given moment at high speeds.

---

local area network (LAN)
The interconnection of computers, hubs, and other network devices in a limited area like a building.

hub
A network connectivity device that connects multiple network nodes together. Used primarily with Ethernet, it forwards all traffic it receives from one port to all other ports.

switch
A network connectivity device that connects multiple network nodes together. Used primarily with Ethernet, it forwards traffic based on the addresses found within that traffic, thus eliminating unnecessary network traffic on other ports.

Access point
A device that acts as a wireless hub, and allows wireless users to connect to a wired network. Also known as an AP.

What Is Networking?

Local area networks have the following characteristics:

- They are used within small areas (such as in an office building).
- They offer high-speed communication—typically, 10Mbps or faster.
- They provide access for many devices.
- They use LAN-specific equipment such as repeaters, hubs, and network interface cards.

NOTE

When measuring how fast a network transmits information, you will typically use one of two different suffixes. Mbps stands for megabits per second, and is equal to one million bits transmitted per second. Kbps is kilobits per second, and is equal to roughly one thousand bits transmitted per second. For more information on bits and bytes, see Chapter 8.

Metropolitan Area Networks

A **metropolitan area network (MAN)** is made up of LANs that are interconnected across a city or metropolitan area. MANs have become increasingly popular as a way of allowing local governments to share valuable resources, communicate with one another, and provide a large-scale private phone service. Although MANs are very expensive to implement, they offer a high-speed alternative to the slower connections found in WANs. MANs offer better speed because of the high-performance cable and equipment used to implement them.

MANs are also appealing to fairly large regional businesses that want to connect their offices. MANs can span as much as 50 to 75 miles, and they provide high-speed network access between sites.

Unlike LANs, in which there are many connections to devices, MANs typically will have just one connection to each site. This is due primarily to the excessive cost of the cable and the equipment. Creating a new MAN connection requires purchasing existing cables from a telecommunications company (the least-expensive option) or having new cables installed, which can cost hundreds of thousands of dollars.

Metropolitan area networks have the following characteristics:

- Sites are dispersed across a city or the surrounding area including the city.
- With the advent of MANs, historically slow connections (56Kbps–1.5Mbps) have given way to communication at hundreds of megabits per second and even gigabit speeds.

bridge
A network device that splits a network into two or more parts for better performance. Information on one part of the network travels through the bridge only if it is intended for the other network.

router
A device used to select the best path for data travel to reach a destination on a different network.

network interface card (NIC)
The internal hardware installed in computers and other devices that allows them to communicate on a network.

metropolitan area network (MAN)
Two or more LANs interconnected over high-speed connections across a city or metropolitan area.

Chapter 1

- They provide single points of connection between each LAN.
- They use devices such as routers, telephone and **ATM switches**, and microwave antennas.

ATM switch
A network device used by telecommunications companies like the local telephone company to support multiple connections on an ATM network.

> **NOTE**
> See Chapter 10 for more about ATM (Asynchronous Transfer Mode).

Wide Area Networks

wide area network (WAN)
Two or more LANs or MANs that are interconnected using relatively slow-speed connections over telephone lines.

A **wide area network (WAN)** interconnects two or more LANs (or MANs) over slow connections leased from the local telephone company. WANs run over telephone cables because they typically cover a wide geographical area—they may span cities, states, or even countries. Interconnecting LANs and MANs over great distances of land and water requires a lot of coordination and sophisticated equipment. In most cases, the telephone company is involved in providing the physical cable connection. When connections are required across the globe, other major telecommunications companies will provide **satellite** connectivity.

satellite
In telecommunications, a device that is sent into earth's orbit to travel around the earth and provide telecommunications services for voice and data.

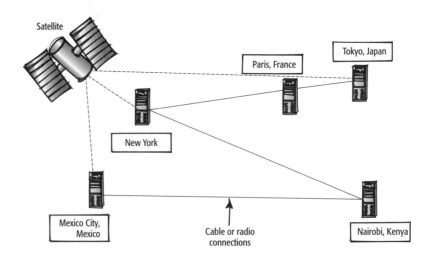

The majority of WANs communicate at speeds between 56Kbps and 1.5Mbps, although speeds up to 45Mbps are available. In fact, the definition of a WAN as a slow-speed connection is changing rapidly. As you will learn in Chapter 14, new technologies are emerging that offer high-speed communication over the telephone network.

What Is Networking?

Wide area networks have the following characteristics:

- They can cover a very large geographical area; even span the world.
- They usually communicate at slow speeds (compared to LANs).
- Access to the WAN is limited—a LAN usually has only one WAN link that is shared by all devices.
- They use devices such as routers, **modems**, and WAN switches.

WAN links are how the Internet was created. By connecting many LANs together using WANs, all connected users are able to share information.

---

modem
A device that turns digital signals to analog, and vice versa, for communication on regular telephone lines.

Chapter 1

Laying the Foundation: The Public Telephone System

public switched telephone network (PSTN)
The telephone infrastructure that relies on circuit switching or other switching technology to open and close circuits for voice conversations.

Federal Communications Commission (FCC)
A government agency that reports directly to the U.S. Congress. The FCC is charged with regulating interstate and international communications by radio, television, wire, satellite, and cable.

The telephone system is the cornerstone of the Internet. When information moves from a website in San Francisco to a computer in Philadelphia, data is transmitted across the **public switched telephone network (PSTN)**. The public switched telephone network comprises several different telecommunications companies that are interconnected, allowing businesses in different locations in the United States to communicate. Although these interconnections have existed for a long time, they were not always available for companies to use as they pleased.

Without the telecommunications laws that guaranteed that all homes would receive telephone service, much of what we know as the Internet would not exist. The infrastructure that is generally referred to as the *telephone system* was made possible by an idea proposed by Theodore Vail, then president of American Telephone and Telegraph Company (AT&T), in 1907. He proposed that AT&T be given a monopoly over the telephone system. He argued that because a home needed to receive telephone service from only one telephone company, it would not be necessary to have more than one telephone company in a given area. The U.S. government would be responsible for fairly regulating AT&T's activities. In exchange, AT&T would provide "universal access"—including service to rural communities.

In 1956, the Hush-a-Phone court case set an important precedent for using the public telephone network. The Hush-a-Phone company had created a device that attached to a telephone to block background noise. The device used no magnetic parts or electricity. AT&T sued Hush-a-Phone, citing that the monopoly granted them by the government prevented anyone from attaching a device to the telephone network. The court ruled in Hush-a-Phone's favor. Ten years later, AT&T sought to block Carter Electronics from attaching its electrical devices to the AT&T telephone network. In the Carterphone case, the **Federal Communications Commission (FCC)** ruled that other devices could be attached to the telephone network. This case and the preceding one are important to internetworking because they granted other companies the right to build and sell telephones and later to attach network devices to AT&T's telephone network.

The AT&T breakup into the Baby Bells and increased competition have helped to further expand the telecommunications infrastructure in the United States and around the world. The continuing deregulation of local telephone companies is furthering high-speed network access to the home at very affordable rates.

What Is Networking?

The Origin of Networks

In the world of computer networking, things change very rapidly. What's interesting is that the underlying technology does not change much at all. The network may be faster or more efficient, or have more functions, but the way it works is essentially the same. Network administrators need to be knowledgeable about the technologies that led to the networks of today. That knowledge may be useful when troubleshooting a problem or when trying to explain why the Internet works the way it does.

Some of the first networks were really just **mainframe** systems that ran over dedicated lines. In some cases, the lines ran long distances. The modern computing network really began to emerge with the development of two networks. The Semi Automatic Ground Environment (SAGE) was one of the early networks (1958) that was developed to link government computers at radar stations in the United States and Canada. In the 1960s, researchers at MIT developed the **Compatible Time-Sharing System (CTSS)** on an IBM mainframe. The time-sharing system allowed multiple users to run tasks concurrently on the same system. CTSS later included modems to connect over dedicated lines in the lab. A user dialed a single number to access the system. Eventually, the system was used campus-wide, and even provided some users access from home.

One of the first commercialized applications that utilized remote access for online transactions was installed for American Airlines in 1964. IBM's SABRE reservation system linked 2,000 machines in 65 cities to two IBM mainframes using telephone lines. These mainframes considered remote terminals to be "remote inquiry stations," and processed their requests. They could deliver information about any flight within three seconds.

Network Communication

The mainframe systems in use in the 1960s and 1970s relied on the mainframe for all processing of information. Terminals attached to the mainframe would wait their turn as information was processed at the mainframe and sent back to each terminal's screen. These early networks were large, expensive, and difficult to use. On mainframe networks, adding an additional terminal was a very difficult and expensive undertaking. Still, these difficulties did not slow user demands. The emerging networks were seen as a way to share resources. They made it possible for more people to connect and communicate. The network could provide time-saving functions like modifying documents without retyping all of the information. Major changes occurred in networking in the early 1970s with the creation of communication protocols.

mainframe
A large, powerful computing machine that stores and processes information, and that runs applications for the terminals that are attached to it.

Compatible Time-Sharing System (CTSS)
Developed in 1961 at MIT, CTSS had a capacity of up to 30 modems to give terminals access to run tasks concurrently. CTSS was the precursor to operating systems such as UNIX.

The first network protocols were **Token Ring**, **ARCNET**, and **Ethernet**. Each protocol used a different method for computers to access the network. The most significant of the three was Ethernet. Robert Metcalf, then a graduate student at Harvard University, first drew the concept for Ethernet on a piece of paper as part of his Ph.D. thesis. The purpose of the research project was to explore packet switching on the **ARPAnet** and **ALOHAnet** networks. Today, Ethernet is the most widely used access method for computer networks.

ARPAnet

While universities and private companies were extending networks using mainframes and terminals, the United States Department of Defense was developing its own network. In the early 1960s, the Advanced Research Projects Agency (ARPA) of the Department of Defense had begun work on a network called ARPAnet. ARPAnet was an experimental network that was created as a communication solution that could withstand a partial failure caused by a bomb attack. ARPAnet also gave top researchers at universities and government institutions the ability to collaborate. After almost a decade in development, the first nodes were connected at the University of California at Los Angeles, the University of California at Santa Barbara, the Stanford Research Institute, and the University of Utah. By 1971, there were 23 nodes connected to ARPAnet. The primary application in use was e-mail.

ARPAnet continued to grow rapidly in the 1980s. By 1989, the ARPAnet was dissolved as a single entity, leaving the public infrastructure known as the Internet and the military system renamed DARPAnet (Defense Advanced Research Projects Agency network). ARPAnet is significant both because it became the Internet and because it demonstrated the ability to interconnect different networks from around the world using the existing public telephone network.

Token Ring
A network access method that relies on tokens to allow devices to transmit information.

ARCNET
A network access method that uses tokens such as Token Ring, but is much less expensive.

Ethernet
A network access method that allows any directly connected device to transmit on the network, provided that no one else is transmitting.

ARPAnet
The predecessor to the Internet, ARPAnet was developed by the Department of Defense's Advanced Research Projects Agency to provide reliable communication, even in the event of a partial network failure.

ALOHAnet
A network that connected the Hawaiian Islands using radio transmissions.

What Is Networking?

The Internet

The Internet grew out of ARPAnet. ARPAnet's unique purpose of providing reliable service, even in the event of a partial failure, proved to be a critical function to the success of the Internet. Another reason for the rapid expansion of ARPAnet and the success of the Internet had to do with the distribution of a not-so-well-known piece of software called BSD UNIX.

By using BSD UNIX, and by funding a company called Bolt, Beranek and Newman, Inc., ARPA was able to expand the Internet to most university computer science departments. BSD UNIX, a computer operating system, proved to be the tool that universities needed to access ARPAnet and communicate with peers.

The Internet became more popular and gained support when the National Science Foundation formed **NSFNET**. NSFNET linked supercomputers at five educational centers, the University of Illinois at Urbana-Champaign, the University of California at San Diego, Princeton University, Cornell University, and Pittsburgh University. These sites soon led to the development of regional networks. NSFNET, with its powerful infrastructure, became the **backbone** of ARPAnet, thereby interconnecting more networks. In 1991, the NSF permitted the first use of the Internet for commercial purposes. By 1995, the popularity of the Internet had exploded, and the NSF decommissioned its own backbone infrastructure, leaving the Internet as a self-supporting industry.

The Future of the Internet

The Internet in use today looks very different than it did in 1990. At that time, there were fewer than 250,000 users on the Internet. It is estimated that there will be more than 300 million users worldwide in just a few years.

For the Internet to be able to handle the growing number of users, major improvements will have to be made in the infrastructure. Even now, the "backbone" of the Internet is being upgraded to the latest network technology. In addition, a second Internet has been built called **Internet 2** (the Abilene Project). Learn more about Internet 2 at `http://www.internet2.edu/`.

Internet 2 is a **very high performance Backbone Network Service (vBNS)**, sponsored by the NSF along with many other government and commercial partners. Internet 2 was developed and installed to support the research and communication needs of academia. To date, there are 95 university campuses, government agencies, and cooperatives connected across the United States, and more are being added. These 95 sites are linked together by twelve regional **gigabit Points of Presence (gigaPOPs)**.

NSFNET
The name for the network backbone funded and built by the National Science Foundation that connected many isolated networks to the ARPAnet.

backbone
The main connection point for multiple networks that carries the bulk of the traffic between different networks.

Internet 2
An internetwork connecting major university campuses, research institutes, and government agencies across the country for research and collaboration.

very high performance Backbone Network Service (vBNS)
The gigabit network developed and managed by MCI in cooperation with the National Science Foundation and other agencies.

gigabit Point of Presence (gigaPOP)
A site that is considered a main backbone provider for Internet 2, and is capable of supporting internetwork speeds in the gigabit range.

Review Questions

Terms to Know
- access point
- ALOHAnet
- ARCNET
- ARPAnet
- ATM switch
- backbone
- bridge
- Compatible Time-Sharing System (CTSS)
- Ethernet
- Federal Communications Commission (FCC)
- gigabit Points of Presence (gigaPOPs)
- hub
- Internet 2
- internetwork

1. Why was AT&T given a monopoly over local phone service?

2. What was the significance of the Carterphone case?

3. What governmental agency developed the ARPAnet?

4. What important invention was created by Robert Metcalf?

5. What are the four main characteristics of LANs?

6. When would a MAN be used?

7. Describe how WANs are different from LANs and MANs.

_____

_____

8. What are two reasons why ARPAnet and later the Internet were successful?

_____

_____

9. What network added to the overall capacity of the Internet, and in what way?

_____

_____

10. What is one of the major accomplishments of Internet 2?

_____

_____

Terms to Know
- local area network (LAN)
- mainframe
- metropolitan area network (MAN)
- modem
- network interface cards (NICs)
- NSFNET
- protocol
- public switched telephone network (PSTN)
- router
- satellite
- switch
- Token Ring
- very high performance Backbone Network Service (vBNS)
- wide area network (WAN)

Chapter 2

Communication Models

Communication, in the most basic sense, is the act of exchanging information with another person. People can communicate in a variety of ways, including conversation, written letters, body language, and electronic mail. But in order to communicate, we need a communication model—a frame of reference or set of guidelines so that we can understand each other.

Computers, like people, also need appropriate models to communicate effectively. Computers use a set of rules that allow them to communicate, even when the computers are running different operating systems or are located on different types of networks. The theoretical model used for communication between devices on a network is the Open Systems Interconnection model, or OSI model. This model is considered the fundamental framework for the way all devices on a network should communicate.

In this chapter, you will learn all about the OSI model. At the end of the chapter, you should be able to answer questions on the following topics:

 The purpose of the OSI model and its seven layers

 The jobs each of the layers performs

 The process for communication between devices

 The four layers of the TCP/IP (or DoD) model compared to the OSI model

Chapter 2

What Is the OSI Model?

OSI (Open Systems Interconnection) model
The communications model developed and adopted by the ISO in 1977. It defines how hardware and software should be developed to support specific functions for communication between devices.

International Organization for Standardization (ISO)
An international standards organization dedicated to defining global communication and standards.

devices
Hardware that is capable of attaching to a network and communicating with other hardware. Examples are computers, printers, routers, repeaters, hubs, and bridges.

The **OSI (Open Systems Interconnection) model** is the foundation for all communications that take place between computers and other networking devices. The **International Organization for Standardization (ISO)** began developing the OSI model in 1974, after the United States Department of Defense (DoD) developed and began using the TCP/IP (Transmission Control Protocol/Internet Protocol) stack. TCP/IP is a suite of protocols that work together to provide communication between network devices. It is important to note that the OSI model is just that: a model. It is not a protocol that can be installed or run on any system. TCP/IP, on the other hand, is a functioning protocol that enables computers to communicate. After many years of discussion, the OSI model was finally adopted in 1977. Today, it is used as the theoretical model for the way communication between **devices**, such as computers, takes place.

The OSI model is important to understand because it is the basis for all network communication models that have been developed and implemented. Although it is a theoretical model, it has defined the way in which devices interact since the late 70s. Understanding the purpose of each layer and its relationship to the other layers will help you design, implement, and, most importantly, troubleshoot any network.

There are seven layers in the OSI model: Application, Presentation, Session, Transport, Network, Data Link, and Physical. These layers make up a framework that defines the way in which physical hardware, media (such as cables), and software work together to communicate.

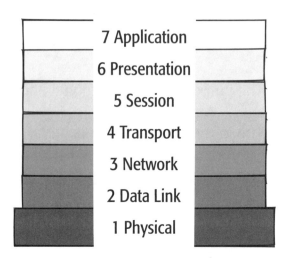

Physical connection

Communication Models

Each layer of the OSI model is independent from every other in its purpose and responsibilities. Each must do its own job, as well as provide the ability for information to move layers above and below it. In this way, the model creates a modular framework for understanding how network communication is taking place, as well as aid in troubleshooting problems that arise within a network.

As you can see in the following illustration, when two devices communicate, each device is responsible for utilizing each of the functions of each layer to ensure that the data will be transferred to the appropriate layer on the receiving end. For instance, the Application layer of node A communicates with the Application layer of node B by passing the data through the other layers. The Application layer of either node is not concerned with the functions of the other layers; as long as the other layers are doing their jobs, the Application layer can send and receive data and determine whether it has an application that can process the data. For example, if you send an HTTP request for a web page from a server that is not running web server software, the Application layer of the server will not be able to respond to the request. The result is that you see an error message on your screen.

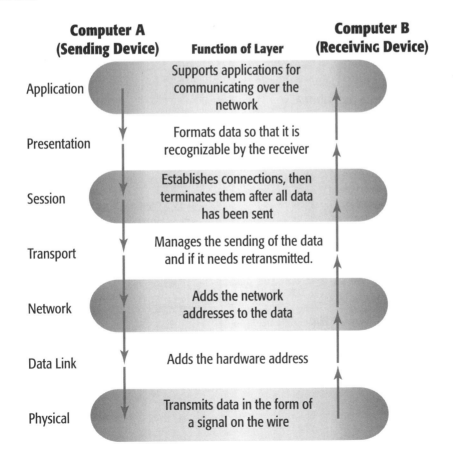

Chapter 2

How Data Flows through the OSI Model

source
The device in which data being sent over a network originates.

data
A term used to represent the information that is transferred and formatted in the top three layers of the OSI model.

packet
Describes the form of data after it has passed through the Transport layer. A packet is a group of bits that includes a header, data, and trailer; and can be transmitted over a network.

destination
A device on a network that is the recipient of data transmitted by the sender.

encapsulated
The process of enclosing a packet of data with information from the current layer as it passes down the OSI layers.

deencapsulated
Layer information is stripped from a packet as it moves up the OSI layers.

The movement of data through the OSI model is easy to follow. When two devices want to communicate with each other, data will be sent from the Application layer of the sending computer or device, referred to as the **source**. The **data**, in the form of a **packet**, will continue down the layers of the OSI model of the source device until it reaches the Physical layer: layer 1. The Physical layer is where the data begins its journey out into the network.

At the Physical layer, the data represented in the computer as bits, or 1s and 0s, is converted into a format for transmission over the network. The information can be transmitted in the form of electricity, light, or radio waves. When the data reaches the **destination** device (the device that the source wants to talk to), it travels up the layers of the OSI model until it reaches the user.

As data flows down the OSI layers of the source computer, it is **encapsulated** with more and more information. The encapsulation process is illustrated as follows. Notice that as the data moves through the OSI layers, header and trailer information are added to the packet. The data inside the packet does not change. At the destination device, a similar process occurs. The information is **decapsulated** and sent up the OSI layers. It is each layer's job to package—or unwrap—the data correctly for its neighbor.

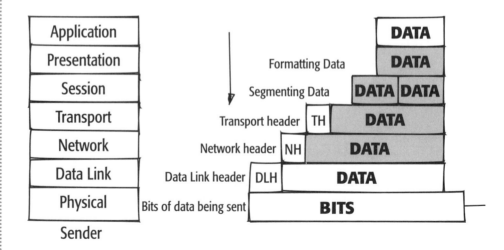

Communication Models

Understanding Each OSI Layer

Part of each layer's job is to perform specific functions for processing the data before it is passed on to the layer above or below. For the sending device, that means that each layer must process the data and prepare it for the layer below it. For example, when the Session layer wants to begin a communication session, it prepares to pass the data and instructs the Transport layer to create the session.

In the receiving device, each layer provides services to the layer below it. These services can be divided into horizontal and vertical communications models. The communication between layers in the same device is an example of the vertical communication model, also called "peer-layer communication." Network administrators, engineers, and software developers have a better understanding of the OSI model when analyzing the communication process horizontally.

At the top of the OSI model is the Application layer—the layer that is the starting and ending point for a communication from a sending device to a receiving device—which is referred to as layer 7. The "bottom" layer, the one that actually sends the message on its way, is layer 1: the Physical layer. This is often confusing for people. To remember the order of the layers, try using a mnemonic device that begins with layer 1, the Physical layer, and ends with layer 7, the Application layer. One common **mnemonic device** is "All People Seem To Need Data Processing."

mnemonic device
A way of remembering information by using a phrase, song, or some other method.

Application	**A**ll
Presentation	**P**eople
Session	**S**eem
Transport	**T**o
Network	**N**eed
Data Link	**D**ata
Physical	**P**rocessing

Layer 7: The Application Layer

The first layer of the OSI model we'll look at is the layer closest to the user, the **Application layer**. As noted earlier in this chapter, this is layer 7 of the OSI reference model. If you have viewed web pages from a web server, then the data was processed by a protocol that functions at the Application layer.

Protocols functioning at the Application layer work with the applications you use to communicate over the network. Examples of protocols that are needed for applications to work on a network are **SMTP (Simple Mail Transfer Protocol)** to send e-mail to a friend, **HTTP (HyperText Transfer Protocol)** to access web pages while you are surfing on the Internet, or **FTP (File Transfer Protocol)** to download a file from an FTP server.

When you use your mouse to click on a link on a website or type in a URL on the Internet, you are giving a command. This action tells the browser to retrieve information from the appropriate web server on the Internet. When you request information like a web page from the Internet, your computer is considered the source and the website's computer is the destination. The Application layer's services are used to complete the job and deliver the information you requested.

The following illustration demonstrates how this process works:

Application layer
The top layer of the OSI model. The Application layer formats data for a particular function such as network printing, electronic mail, or web viewing.

Simple Mail Transfer Protocol (SMTP)
A protocol that specifies how electronic mail is to be delivered to its destination.

HyperText Transfer Protocol (HTTP)
A protocol that defines how web pages are processed by web servers.

File Transfer Protocol (FTP)
A protocol that defines how files are transferred to and from an FTP server and a client computer.

Communication Models

Application Layer Services

There are literally hundreds of services that function at the Application layer. It is a common mistake to equate the purpose of the Application layer with computer applications or programs. Business productivity software, for example, does not inherently run at the Application layer, although the software may make use of network services.

File Services
Any application that provides network users to files and folders on a computer or server.

There are five categories of services that are very common to most Internet and network users. These categories are as follows:

File services The most common of the services provided by the Application layer are **file services**. File services may include requesting a web file to view a web page, as in the previous example, or accessing a shared file on a local network.

> **NOTE**
>
> Remember that a file does not have to be a text document. File types can include graphic files such as JPEG, GIF, or PICT files; and videos in Quick-Time or MP3 format. These and other files, regardless of file format, are transmitted using some type of file service that operates at the Application layer.

Electronic-mail services E-mail servers rely on the SMTP protocol to pass e-mail messages to each other. SMTP is the protocol that allows applications such as Sendmail in UNIX, Exchange in Windows, and other e-mail server software to transfer messages to one another.

Network-printing services Unlike printers that are directly attached to the computer, networked printers are used in situations in which it is not economical to give each user a dedicated printer. Networked printers are created by using either hardware- or software-based print servers. The printer servers use various protocols to deliver print capabilities to the clients over the network.

Application services Networked applications were common in the first decades of networking. Computer users relied on servers to give them access to applications. Due to the high cost of hard drives, applications running on the user's computer weren't even a consideration. Although personal computers are much more powerful today, there are still many instances when a networked application makes more sense.

Database services One of the most common types of application services in use is a database service. Databases are used in everything that requires the storage and retrieval of information. On a network, database services use a number of different upper-layer protocols to input and extract data in the database.

Chapter 2

NOTE

For more information on servers and the network services they provide, see Chapter 4.

Presentation layer
Layer 6 of the OSI model. The Presentation layer manages the conversion from data structures used by the computer to a form necessary for communication over the network.

Layer 6: The Presentation Layer

Below the Application layer is the **Presentation layer**. This is the sixth layer of the OSI model, one layer away from the end user. The Presentation layer has three main jobs:

- Data presentation
- Data compression
- Data encryption

abstract syntax
The format of data in the Application layer of the OSI model before it is converted to any other format by the Presentation layer.

The first function, data presentation, ensures that the data being sent to the recipient is in a format that the recipient can process. This function is important because it enables the receiving device to understand the information from the sending device. The sending device performs the data presentation service by turning the request from its own native format (its **abstract syntax**) into a common language (**transfer syntax**). For most computers, that common language is ASCII (American Standard Code for Information Interchange). ASCII uses 96 letters and numbers as well as 32 nonprinting characters.

transfer syntax
The format of data after it has been converted by the Presentation layer into a "common language" format, typically ASCII.

If you were requesting services from an IBM mainframe, the common language the Presentation layer would translate your request into would be EBCDIC (Extended Binary Coded Decimal Interchange Code). EBCDIC uses 256 special characters.

data compression
The reformatting of data to make it smaller.

The second function of the Presentation layer, **data compression**, shrinks large amounts of data into smaller pieces. This allows data to be transferred more quickly across a network.

NOTE

In networking, data is often referred to as a *packet*. Packet is the term used to identify data as it moves from device to device on a network.

Communication Models

In the next example, the source computer has a simple text document to send to the destination computer. At the Application layer, this data looks like a normal sentence with formatting intact. Before the data can be sent to the destination computer, the text passes through the OSI layers. When it reaches the Presentation layer, the text data is compressed, transformed, and possibly encrypted; then sent on through the remaining layers. At the destination computer, when the information reaches the Presentation layer, it is returned to its original format (with spaces and formatting), and then sent on to the Application layer.

encryption
The process of converting data into a random set of characters that is unrecognizable to everyone except the intended recipient.

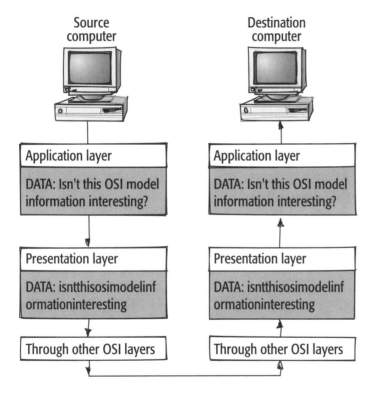

The Presentation layer is also responsible for data **encryption**. Encryption is an important tool that allows us to hide information from everyone except the person who originally sent the information and the intended recipient.

NOTE

Not all messages sent across a network are or need to be encrypted Today, with the use of credit cards on the Internet, the prevalence of hackers, and the need for personal privacy, encryption of some sort has become almost a necessity.

When the encrypted message or data arrives at the destination device's Presentation layer, the encryption service of the Presentation layer is responsible for decrypting the message. The destination device or its user must have the "key" to unlock the code. This is often in the form of a password that the user supplies, but may also include the use of a digital key that is created using software.

Layer 5: The Session Layer

Following down the model, the **Session layer** has the primary responsibility of beginning, maintaining, and ending the communication between two devices, which is called a **session**. It also provides for orderly communication between devices by regulating the flow of data. The services that the Session layer is responsible for include the following:

- Establishing a connection
- Maintaining the session
- Ending the connection
- Dialog control
- Dialog separation

Establishing, Maintaining, and Ending a Session

When a sending device first contacts a receiving device, it sends a **syn** (synchronization) packet to establish communication and determine the order in which information will be sent. The receiver sends back what is called an acknowledgment, or **ack**. This is how the receiving computer acknowledges that there is a request to begin a communication session. The session can then be set up. At the same time the receiver sends the acknowledgment, it also sends a syn packet; this is necessary for two-way communication to occur. At the end of the session, the sender transmits an acknowledgment to request an end to the session. When the communication session ends, protocols at the Session layer are responsible for its termination.

The Session layer of the receiving device works with the Session layer of the sending device to set up the rules of communication. The Session layers can make decisions about the conversation based on the abilities of both devices, limiting the number of packets sent and deciding who is going to transmit first.

Once a link is established, it must be maintained for as long as the communication is taking place. One way the Session layer does this is through **keep-alive messages**. These special messages maintain a connection between devices, even when there is no data being transferred.

Session layer
Layer 5 of the OSI model, the Session layer creates, maintains, and terminates communication between devices on a network.

session
A communication channel that is created and maintained between two networked devices in order to transfer data.

syn (synchronization)
The message used to establish communication between two or more systems.

ack (acknowledgment)
The message used to respond positively to a request for synchronization when establishing a communication session between two devices.

keep-alive message
A data packet sent between Session layers to keep inactivity from causing the connection to close down.

Communication Models

After a communication has ended, it is the responsibility of the Session layer to end the conversation. This guarantees that the device with the open session is not sending erroneous data on the network by continuing to send packets to the other device.

Dialog Control

When a device is contacted, the Session layer is responsible for determining which device participating in the communication will transmit at a given time, as well as controlling the amount of data that can be sent in a transmission. This is called **dialog control**. The types of dialog control that can take place include simplex, half duplex, and full duplex.

dialog control
A function of the Session layer that determines which device will communicate first and the amount of data that will be sent.

dialog separation
The use of markers within the data to determine whether all information received is intact.

NOTE

See Chapter 7 for more details about simplex, half-duplex, and full-duplex dialog control.

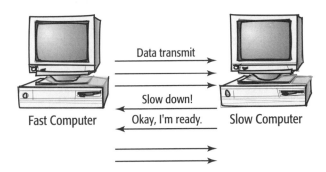

Dialog Separation

The Session layer is also responsible for inserting markers within messages. The process of inserting markers, known as **dialog separation**, ensures that if there is a loss of packets or any other problems during transmission, the conversation can continue. The Session layer knows the data it needs to retransmit because of the checkpoints it has placed within the message.

NOTE

In addition to dialog separation and dialog control, the Session layer is responsible for as many as 20 other services, most of which are used for exchanging information between systems.

Layer 4: The Transport Layer

The next layer of the OSI reference model is the **Transport layer**. The primary function of the Transport layer is to ensure that the data packets are sent and, depending on the network, that the data packets are received intact. The Transport layer does this by using two types of transmission methods: connection-oriented and connectionless.

The Transport layer also has the job of managing the speed of communication between devices. This is known as flow control.

Connection-Oriented Transmissions

Connection-oriented transmissions take place when the receiving device sends an acknowledgment, or ack, back to the source after a packet or group of packets is received. This type of transmission is known as a *reliable transport method*. Because connection-oriented transmission requires that more packets be sent across the network, it has often been considered a slower transmission method, although today's faster networks have made that argument a moot point.

The features of connection-oriented transmission are the following:

- Reliability
- Slower communication
- Packets are resent if a packet is unrecognizable or is not received

If there are problems with the data that is transmitted, the destination computer requests that the source resend the data by acknowledging only the packets that have been received and are recognizable. Once the destination computer receives all of the data necessary to reassemble the packet, the Transport layer will assemble the data in the correct sequence and then pass it up to the Session layer for processing.

Connectionless Transmissions

Connectionless transmissions, on the other hand, do not require the receiver to acknowledge receipt of a packet. Instead, the sending device assumes that the packet arrived just fine. This approach allows for much faster communications between devices. The trade-off is that connectionless transmission is less reliable than connection-oriented transmission.

Transport layer
Layer 4 of the OSI model, the Transport layer takes the data from the Session layer and breaks it up into segments that can easily be transmitted by the lower-layer hardware.

connection-oriented
During communication between two devices, the receiving device will acknowledge to the sender that it has received the data. If part of the data is not received, the sender retransmits the data.

connectionless
When communication occurs between devices, there is no acknowledgment that data has been received.

Communication Models

The features of connectionless transmission are as follows:

- Little or no reliability
- Faster transmission
- Packets are not retransmitted

Connection-Oriented Transmission

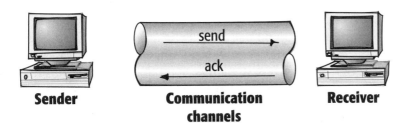

Connectionless Transmission

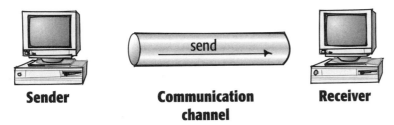

flow control
A Transport layer feature that manages the flow of data. If the receiving device is unable to process incoming data, it sends a message to halt the transmission; when able to continue, it tells the sender to begin transmitting again.

Flow Control

Another responsibility of the Transport layer is to make sure that the sender and receiver communicate at a rate they both can handle. This is called **flow control**. Flow control keeps the source from sending data packets faster than the destination can handle.

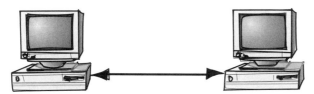

Transmission must occur at agreed-upon speed— like 5 packets at a time instead of 20.

No matter which type of transmission is used, the Transport layer will determine the largest size packet that can be sent and proceed to divide the data in preparation for sending. The Transport layer is also responsible for numbering these chunks so that the destination computer can reassemble them in the right order.

Layer 3: The Network Layer

Below the Transport layer is the **Network layer**, layer 3. This layer is responsible for the addressing and delivery of packets, also known as **datagrams**. The address that the Network layer uses—the **network address**—is a logical address. A logical address is still a number, but the number has no relationship to a permanent physical address or to the hardware address on the network card.

To do its job, the Network layer must do the following:

- Add the network address to the packet (through encapsulation). With some network technologies, network addresses are assigned by the system. In the case of the Internet, network addresses are assigned to local devices by the network administrator; assigned dynamically by a special server on the network; or, if a dial-up account to an Internet service provider is being used, received from the ISP once the user logs on. The network addresses available for use are received from **InterNIC**, RIPE (the European equivalent of InterNIC), and APNIC (the Asia-Pacific equivalent of InterNIC).
- Map the network address to the device's physical address.
- Determine the best path for the packet, based on information it has about the network (this is called routing).
- Ensure that the packet is in the correct format for the destination network.

> **Network layer**
> Layer 3 of the OSI model, the Network layer takes the packets passed down from the Transport layer, and adds the appropriate network addresses to them.
>
> **datagram**
> A data packet that has had the network address added to it as part of the encapsulation process.
>
> **network address**
> The address assigned by the network administrator, based on the network where the device is located. Also known as the *logical address*.
>
> **InterNIC**
> The organization responsible for managing the allocation of IP addresses and domain names for all organizations.

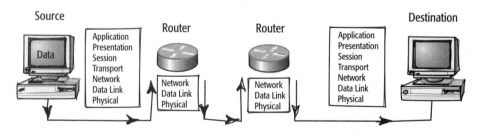

Communication Models

How It Works

When a packet arrives at the Network layer, the Network layer adds source and destination addresses through a process known as **encapsulation**. Both addresses are logical. The first address that is added to the datagram is the source address of the computer that is sending the data. Then the destination address, again a logical network address, is added for the recipient computer. The network addresses are necessary for packets to be transmitted between **end systems**. These devices may exist on either a local network or on other distant networks.

Before a packet can be sent to the destination device, the best path to that device must be determined. For the Network layer to provide this service, called **routing**, the source computer must address the packet using the network address of the destination device. The network address allows the Network layer to then make decisions about how to route the packet to the appropriate network.

When the packet is sent, it may need to "switch" networks, again using routing. The protocols operating at the Network layer provide routing services by acting like an air traffic controller or traffic cop rerouting the packet to the next segment of its journey.

Routing decisions are usually determined by calculating the shortest path needed to reach the destination, although other factors may apply. At the Network layer, routing decisions are made by routing protocols, which are implemented in software. In most cases, networks use dedicated devices called routers to perform this function.

The Network layer also has the job of making sure the data packet is in the correct format for the type of network it is entering. For example, if one network's media allows for longer packet lengths, the Network layer will format the packets so that the network they are entering can handle them.

A router is the primary piece of hardware that works at the Network layer. All the devices—such as routers—that a packet passes through on its way to the destination device are called **intermediate systems**. These intermediate systems need to deal with the packet only up to the Network layer of the OSI model. They do not go any higher up the stack.

encapsulation
In LAN protocols, the process of adding network address information to a data packet. When the data packet has been encapsulated, it is called a *datagram*.

end system
A device participating in a communication with another device.

routing
The use of a routing protocol to select the best path to the appropriate network.

intermediate system
Any device that is used to assist in transporting data between two communicating end systems.

NOTE

For more information on routing protocols and devices, see Chapter 11.

Layer 2: The Data Link Layer

The next layer of the OSI model is the **Data Link layer**. The Data Link layer has two sublayers, and each provides its own services in the OSI model. The first is the **Logical Link Control (LLC)** sublayer, and the second is the **Media Access Control (MAC)** sublayer.

The LLC sublayer provides the interface between the media-access method and Network layer protocols such as the Internet Protocol, which is a part of the TCP/IP protocol suite. In contrast, the MAC sublayer is responsible for the connection to the physical **media**. Other Data Link layer functions include adding a device's physical address (MAC address) to a datagram and converting the datagrams into frames to be transmitted over the network media.

The LLC sublayer uses two types of connection services that bridge the MAC sublayer to the upper-layer protocols: connectionless and connection-oriented. As in the Transport layer, a connectionless service assumes that data has arrived correctly at the destination. The connectionless service, or LLC type 1, is most common in today's networks because the need for a connection-oriented service is taken care of at the Transport layer.

Connection-oriented service, LLC type 2, checks that a message arrives correctly. Although it may seem that having this reliability is better, the extra work and time involved lowers the performance of the network, especially when this service is already provided at the Transport layer.

> **NOTE**
> The processes of the LLC sublayer are described in the IEEE 802.2 standard, which is discussed in more detail in Chapter 10.

Data Link layer
Layer 2 of the OSI model, the Data Link layer receives data from the Network layer, and packages it as frames to be sent onto the network by the Physical layer.

Logical Link Control (LLC)
The LLC sublayer establishes whether communication is going to be connectionless or connection-oriented at the Data Link layer.

Media Access Control (MAC)
The MAC sublayer adds the destination and source physical addresses to a frame before sending it on to the Physical layer.

media
The means that allow communication to take place—cable, for example.

Data Link layer

LLC sublayer:
Connection-oriented?
Connectionless?

MAC sublayer:
Adds MAC address

Communication Models

At the MAC sublayer of the Data Link layer, the actual physical address of the device, called the **MAC address**, is added to the packet as the encapsulation process continues. The packet, now referred to as a *frame*, at last has all the addressing information necessary to travel from the source device to the destination device. Without both the physical MAC address given to it at this layer and the logical network address given to it at the Network layer, a data packet will not arrive at its final destination.

Header	Destination MAC	Source MAC	Destination address	Source address	LLC header	Data	CRC	Trailer

A device's MAC address is located on its **network interface card (NIC)**. The MAC address is permanently hard-coded on the card by the manufacturer. It is a unique **hexadecimal** address that is not duplicated anywhere in the world. It consists of six pairs of hexadecimal digits; the first six digits are assigned to the NIC manufacturer and the last six are unique. This is truly the physical address of the device.

MAC address
The address of the device that is found on the NIC. Also known as the *physical address*.

network interface card (NIC)
An electrical device installed in a computer that allows it to be connected to a network.

hexadecimal
A numerical system that uses the first six letters of the alphabet to extend the possible digits to 16 beyond the 10 available in the decimal system.

TIP
Find out the MAC address of your Windows 98 computer machine by typing **winipcfg** at the Run or DOS prompt. It is listed as the *hardware* or *adapter* address. You can accomplish the same task by using the **ipconfig** command at the same prompt in Windows NT/2000/Me/XP.

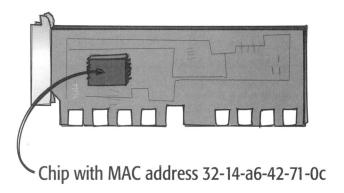

Chip with MAC address 32-14-a6-42-71-0c

NOTE
Cisco uses a dotted-hexadecimal notation format when identifying MAC addresses. It consists of three sets of two pairs each of hex digits. Thus, the previous MAC address would look like this: 3214.a642.710c.

Chapter 2

Layer 1: The Physical Layer

Below the Data Link layer is the **Physical layer**—the bottom layer or layer 1 of the OSI model. The Physical layer is responsible for the actual physical connection between devices. This physical connection may be made using a variety of materials such as the following:

- Twisted-pair cable
- Fiber-optic cable
- Coaxial cable
- Wireless communications

The same NIC that is involved in services at the Data Link layer also helps at the Physical layer. The NIC is responsible for converting the data, called *bits*, into transmission signals. The signals generated by the NIC may be in a variety of formats, depending on the network connection medium. These transmissions may be **analog** or **digital**, though both types transmit **binary** data. The Physical layer is also responsible for the rate at which these transmissions are sent.

Another function of the Physical layer is that it manages the way a device connects to the network media. For example, if the physical connection from the device to the network uses coaxial cable, the hardware that functions at the Physical layer will be designed for that specific type of network. All components, including the connectors, will be specified at the Physical layer as well.

Physical layer
The bottom layer of the OSI model, the Physical layer specifies the type of media to be used, the transmission format, and the topology of the network.

analog
A type of signal that uses a continuous waveform combining amplitude, frequency, and phase.

digital
A type of signal that uses pulses to send binary signals across media. One pulse represents a 1; a lack of a pulse represents a 0.

binary
A number system in which information is represented as either a 1 or 0.

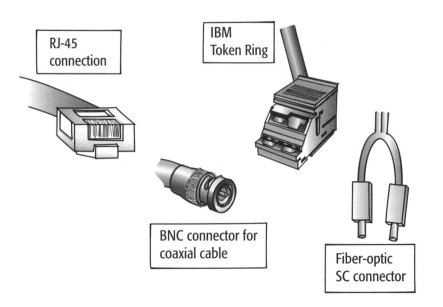

Communication Models

There are many devices that function specifically at the Physical layer. These include NICs, repeaters, hubs, and concentrators. These devices can regenerate the signals produced by the NIC as they travel through the network. If they will be moving onto a different type of network that requires a different type of signal or frame format, the devices can make changes to the format.

> **NOTE**
>
> For more information on NICs, repeaters, hubs, and concentrators, see Chapter 9.

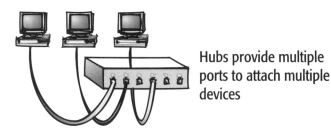

Hubs provide multiple ports to attach multiple devices

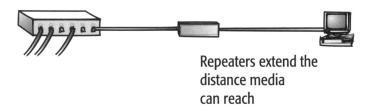

Repeaters extend the distance media can reach

The Physical layer also determines the type of topology used by the network. You'll learn more about the physical topologies used in networking in Chapter 6.

33

The OSI Model versus the TCP/IP Model

TCP/IP protocol suite
Represents several different protocols that may run in tandem with, independently of, or in place of Transmission Control Protocol (TCP) and Internet Protocol (IP).

DoD model
The term used to describe the conceptual model for the TCP/IP protocol suite.

The OSI model is important to understand because it is the modern conceptual model for computer communication. A conceptual or theoretical model like the OSI is not the same as a functioning protocol stack—a combination of protocols that work together.

The most common communications protocol stack on today's networks is the **TCP/IP protocol suite**. Although the TCP/IP protocol suite has been around for decades, it has become widely used because of the growth and popularity of the Internet, which is built on TCP/IP. TCP/IP uses interacting protocols to provide communication between end systems. The model for this protocol suite is known as the **DoD model**, primarily due to the fact that the DoD (U.S. Department of Defense) funded the original project from which TCP/IP was developed.

Unlike the OSI model, in which layers are distinct from one another, the DoD model is a working model because it was used as the basis for the development of the TCP/IP protocol suite. For this reason, the TCP/IP suite of protocols is mapped to the more contemporary OSI model. Regardless, it is possible to compare the functions of the layers of the DoD model to that of the OSI model. From the comparison, a four-layer DoD model can be derived; in some reference materials, a fifth layer, the Physical layer, is added.

OSI Model	TCP/IP Model
Layer 7: Application layer	
Layer 6: Presentation layer	Application layer
Layer 5: Session layer	
Layer 4: Transport layer	Transport layer
Layer 3: Network layer	Internet layer
Layer 2: Data Link layer	Data Link layer
Layer 1: Physical layer	

Application or Process Layer

The Application layer of the TCP/IP model, also known as the *Process layer*, handles the way applications at both the source and destination devices process

Communication Models

information as it is sent and received. There are many application-level services that are included as part of the TCP/IP protocol suite, some of which have been mentioned earlier. These services are discussed in detail in later chapters. The Application layer of the DoD model maps to the Application layer, Presentation layer, and most of the Session layer of the OSI model.

Host-to-Host or Transport Layer

Like the OSI model's Transport layer, the DoD's host-to-host layer manages the flow of data between devices and the type of transmission—connection-oriented or connectionless. TCP/IP uses two protocols to provide this service: TCP (Transmission Control Protocol), which provides reliable connection-oriented service, and UDP (User Datagram Protocol), which provides unreliable connectionless service. More details will be provided later in Chapter 11.

Internet Layer

Although the Internet layer consists of several protocols, the primary protocol referred to at this layer is the Internet Protocol (IP). The Internet Protocol serves several functions, the foremost of which is to provide a hierarchical addressing scheme to identify devices on the network. Hierarchical addressing schemes organize numbers into groups that can be allocated to companies, schools, government organizations, and other organizations, so that every device on the Internet has a unique address. The Internet layer is also significant because it relies on the hierarchical addressing of IP to route data independently of the type of network media.

NOTE

You'll sometimes hear the Internet layer called the *Network layer* or the *Internetwork layer*. None of these names is right or wrong. Use whichever name makes remembering the layer easiest for you.

Network Interface Layer

The *Network Interface layer* manages the transmission of data within a network. After the Internet layer routes the packets to the correct network, this layer makes sure that the data is sent to the correct device.

Physical Layer

Although some models and overviews of TCP/IP include a Physical layer as part of the stack, in reality it is not defined as a part of the DoD model. TCP/IP leaves the physical connection to manage itself. The connected media and the architecture they are servicing are obvious to the Data Link layer.

NOTE

For an in-depth discussion of protocols, see Chapter 11.

Review Questions

1. What is the purpose of the Open Systems Interconnection (OSI) model?

2. What are the seven layers of the OSI Model?

3. How do the layers work together to provide complete intercommunication?

4. Name at least one purpose or job provided by each layer.

5. Which layer is responsible for the setup and breakdown of a communication?

6. Which layer is responsible for end-to-end routing of the data?

Terms to Know
- ❏ abstract syntax
- ❏ ack
- ❏ analog
- ❏ Application layer
- ❏ binary
- ❏ connectionless
- ❏ connection-oriented
- ❏ data
- ❏ data compression
- ❏ Data Link layer
- ❏ datagram
- ❏ decapsulated
- ❏ destination
- ❏ device
- ❏ dialog control
- ❏ dialog separation
- ❏ digital
- ❏ DoD model

Review Questions

Terms to Know
- encapsulated
- encapsulation
- encryption
- end system
- file services
- flow control
- FTP (File Transfer Protocol)
- hexadecimal
- HTTP (HyperText Transfer Protocol)
- intermediate systems
- International Organization for Standardization (ISO)
- InterNIC
- interoperability
- keep-alive messages
- Logical Link Control (LLC)
- MAC address
- media

7. At which layer is the physical address of the device added to the data packet?

8. What is the physical address called, and how is its address assigned?

9. What is the address that is added at the Network layer?

10. What is the process of adding an address called?

11. What are the two sublayers of the Data Link layer?

12. What are the two types of communications used on a network?

13. Name three types of physical connections (media) that can be used in networks.

   _____

   _____

14. Name at least two devices that function at the Physical layer.

   _____

   _____

Terms to Know
- Media Access Control (MAC)
- mnemonic device
- network address
- network interface card (NIC)
- Network layer
- OSI (Open Systems Interconnection) model
- packet
- Physical layer
- Presentation layer
- routing
- Session layer
- SMTP (Simple Mail Transfer Protocol)
- source
- syn
- TCP/IP protocol suite
- transfer syntax
- Transport layer

Chapter 3

Network Architectures

Designing a network to enable computers to communicate is much like designing a building. An architect drawing up the plans for a new office building must consider many factors, including the amount and type of space needed for employees; and structural needs such as electricity, ventilation, and all of the systems that will allow business people to function efficiently. The construction company follows the architect's blueprint, which specifies both the physical location and types of materials needed to make the building complete.

Just as it is essential to come up with a blueprint before constructing a building, it is equally important to decide on a network architecture before setting up a network. The technologies that make up a network are not of much good use separately. People who want to use a network to solve a problem or improve a process need the different components to function as one. A network architecture provides the blueprint for building a network.

In this chapter, you will learn about the following topics:

   Selecting a network architecture

   Peer-to-peer networks

   Client-server networks

   Hybrid networks

Understanding Network Architectures

network architecture
A design that reflects the intended use of hardware and software that will allow computing devices to communicate with each other.

peer-to-peer network
A network architecture that allows 10 or fewer users to effectively share files and folders on their computers with other users on their networks.

client-server network
A network architecture that combines the processing of both the workstation and the server to perform a task.

workstation
Any computer that is capable of processing and storing user data for work-related tasks.

server
A computer used to centralize resources, administration, and security for a network and its users by using specialized hardware and software.

hybrid
On a local area network, the use of both peer-to-peer and client-server network architectures.

network administrator
A person whose job it is to manage and support the network infrastructure, including supporting the needs of users.

As software development has improved over the past several decades, so have computers become better at interacting on a network. The way in which computers interact on a network is known as the **network architecture**. There are three types of network architectures that you need to understand: peer-to-peer, client-server, and hybrid. The type of architecture appropriate for an organization depends on a number of factors, including geographical location, the number of users, any special application needs, and the amount of technical support available.

In peer-to-peer networks, people share their computers' resources, making them available to others. **Peer-to-peer networks** were available as early as 1984, when Apple Computer unveiled its Macintosh Plus. Microsoft introduced peer-to-peer networking capabilities in 1992 with the release of Windows for Workgroups 3.11.

The most basic peer-to-peer network allows people to share resources such as folders, printers, and CD-ROM drives. What does this mean in practical terms? Users can avoid the frustration of relying on floppy disks to share files or print on someone else's printer. Instead, peer-to-peer networking lets a user access a file or printer across the network. File and print sharing on a network saves time, which means employees can be more productive, thereby saving companies money.

Companies and other organizations have found additional cost savings in client-server networks. Client-server network architectures centralize resources and data on servers. The centralized structure of client-server networks offers the additional advantage of added security. This is because it is easier to secure important information when it is located in one place. And centralization makes it easier to protect information in the event of a fire or other disaster. Besides centralized services and security, client-server architectures offer a performance improvement over peer-to-peer networks. When many users need access to the same computer or printer on a peer-to-peer network, performance suffers. Client-server architectures avoid this problem. **Client-server networks** take advantage of the processing capabilities of **workstations**, along with the power of **servers**. In a client-server network, both computers process tasks and information simultaneously. This capability improves performance and efficiency, especially for computing tasks such as printing or searching a database.

In large organizations in which client-server network architectures are implemented, it may be difficult for small groups of people to work efficiently. **Hybrid** network architectures allow client-server and peer-to-peer networks to coexist on the same network. Small groups of users can easily share files and other resources without requiring the intervention of the **network administrator**.

Network Architectures

Peer-to-Peer Networks

Peer-to-peer networking enables users to share resources, files, and printers in a decentralized way. Specifically, peer-to-peer networks have the following characteristics:

- They allow users to share many resources on their computer, including files and printers.
- They're better for groups of 10 users or fewer.
- They're decentralized—user files are not stored in a central location.
- They allow computers to communicate easily.

Peer-to-peer networking is usually an integrated part of the operating system software on your computer. After the software is configured on each workstation, users are responsible for making their specific information available to others (sharing it) and managing the access to that information.

workgroup
The description of a logical group of users organized by job type.

security
In data processing, the ability to protect data from unauthorized access, theft, or damage.

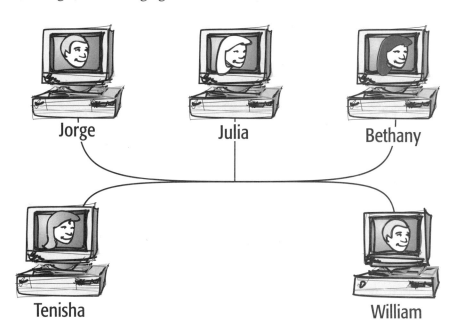

When users are participating in the same peer network, they belong to a **workgroup**. The workgroup is assigned a name, which may represent users who work in the marketing department or the group of people who sit in the cubicles in the northeast corner of the building. The workgroup name makes it easier to remember who is participating in the peer-to-peer network. This is very important in situations in which there may be multiple workgroups on the same network.

Chapter 3

Selecting Peer-to-Peer Networks

Before you decide to implement a peer-to-peer network, you should research your current situation. You may have a business that does not have a network. The employees may have little or no experience with using a network. If this is the case, you will want to keep things as simple as possible. Users will need to know that you are providing them with a tool to help make their work easier. If using the network is complicated or confusing, then people won't use it.

share-level security
The configuration of a shared folder with access permissions using only a password.

The basic criteria for selecting a peer-to-peer network are:

- Ten or fewer users will be sharing resources.
- No server is available.
- Nobody has the time or knowledge to act as a network administrator.
- There's little or no concern about **security**.

The advantages of using a peer-to-peer network are:

- They're easy to configure.
- They don't require additional server hardware and software.
- Users can manage their own resources.
- They don't require a network administrator.
- They reduce total cost.

NOTE

The total cost of a system can be difficult to calculate. The total cost of ownership (TCO) includes the cost of equipment and software, as well as the cost for managing the technology. The management cost includes the time that users spend managing the technology. Look out for these hidden costs that are often overlooked.

The disadvantages of using a peer-to-peer network are:

- They provide a limited number of connections for shared resources.
- Computers with shared resources suffer from sluggish performance.
- They don't allow for central management.
- They don't provide a central location for file saving.
- Users are responsible for managing resources.
- They offer very poor security.

Network Architectures

Security on Peer-to-Peer Networks

Security within workgroups is anything but secure. Peer-to-peer networks use **share-level security**. Share-level security gives the users the authority to assign **passwords** to the local resources on their computers.

Share-level security allows a person who is sharing a resource to implement security with a password, or they can let anyone on the network use the resource by sharing without assigning a password. In some older operating systems, assigning a password to protect the resource required that the password be given to all the people who needed to be able to access that resource. As you can guess, passwords that are shared do not stay secret, or secure, for very long.

The same password could be assigned to every resource, but this would be like a building manager creating the same key for every apartment in their building—not a very good idea in terms of safety. The following illustration demonstrates a better security scheme: a different key is created for each apartment. In peer-to-peer networking, assigning each resource a unique password minimizes the security risks. The drawback, however, is that managing the passwords can be significantly more difficult.

password
A combination of letters, numbers, and symbols either assigned by the administrator or selected by the user that is used in conjunction with the username to gain security access to a resource.

port
Also called a *jack* or *outlet*, a port is a location in which a cable can connect or plug in to the network.

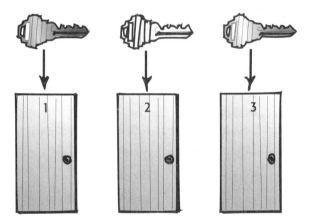

More modern operating systems do a better job at security. Windows 2000 and XP Professional, Macintosh OS 9, and OS X all have a user-management feature that allows you to assign separate usernames and passwords for each individual. But once again, users and their passwords are being created and managed individually on each workstation. This decentralized model of a peer-to-peer network makes for a cumbersome system.

Chapter 3

Implementing a Peer-to-Peer Network

Peer-to-peer networks are much easier to implement than other network architectures. For that reason, they also are more cost-effective. Employees of small businesses who need to collaborate more effectively and efficiently can do so by setting up peer-to-peer services on their workstations.

serial port
A physical interface on a computer that is commonly used to attach a mouse, printer, or modem. A common type of serial interface is RS-232.

Before you can actually build a peer-to-peer network, you will need a shopping list. In addition to the computers, you will need:

- Network interface cards installed in each computer, so that the computers can communicate on the network
- Cable media to connect the computers together
- A device called a *hub*, which has **ports** for connecting the cable to the computers
- Software (included with your OS) that will allow you to share your printers, fax modems, and files

10BaseT
A type of network that specifies the use of a baseband transmission on an 802.3 Ethernet network using unshielded twisted pair and able to transmit up to 10 Megabits per second (Mbps).

NOTE

Network interface cards are discussed in Chapter 2. You'll also learn about network media in Chapter 8.

Making the Network Connections

Creating your peer-to-peer network requires special hardware. **100BASE-T** is a network technology that is reliable, fast, and relatively cheap to install. You will need the following items to build your 100BaseT peer network:

- Network interface card (NIC)—one for every computer
- Hub—enough ports for each computer
- Cable—one for each computer to connect to the hub

Although you can buy each of the pieces separately, many companies offer a "workgroup" or starter kit package to help you quickly build your network. As shown in the following illustration, common components of this type of kit include:

- At least two NICs
- A four- or eight-port hub
- Two patch cables

Network Architectures

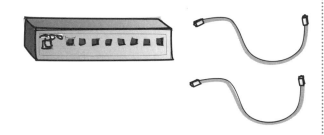

PCI (Peripheral Component Interconnect)
A motherboard technology developed by Intel that improves the speed of communication between an expansion device and the CPU to take advantage of faster CPU speeds.

NOTE

You will learn more about 100BaseT in Chapter 12.

ISA (Industry Standard Architecture)
A motherboard technology that is used to connect expansion cards (NICs, sound cards). Much slower than PCI, ISA was originally designed for the IBM PC and was used in the AT, 386, and 486 models.

When selecting the NIC, you will need to be careful that you are buying the right type for your computer. Newer computers use a type of slot called **PCI**, whereas older computers have **ISA** slots for use with expansion cards or NICs. If your computer is more than three years old, you should check the manufacturer's documentation. You should also be aware that more recent models of personal computers come with a built-in network interface card that is embedded on the motherboard. They work just as well as expansion cards, so there shouldn't be a need to replace the internal cards.

WARNING

Always read the instructions and the safety warnings before working with electronic equipment. Electronic equipment is very sensitive and can easily be damaged.

wrist strap
Also known as an *electrostatic wrist guard*, the wrist strap is worn around the wrist with a cable that connects to the metal case of the computer. Any static electricity from your body is then "grounded," preventing damage to the electronics equipment.

If you know you are going to be installing the NICs in the computer yourself, you should also purchase a **wrist strap**, which will protect your computer from the static electricity coming off your body.

After you have properly installed the network cards, you can connect each end of the cable to the computer and the hub. Next, you will need to configure the software on your computers.

47

Chapter 3

Configuring Windows Peer-to-Peer Networking

Windows 95
An operating system developed by Microsoft to run on PC-compatible computers. Released in 1995, it unveiled a more-friendly GUI, enhancements for networking, and a hardware auto-detect feature called Plug and Play.

Windows 98
A Microsoft operating system that retains the essential parts of Windows 95, but with new applications for Internet use and improved hardware recognition.

Windows NT Workstation 4
Microsoft's premiere desktop operating system, NT Workstation is essentially Windows NT Server. It has fewer network management applications and services than the server version, and each workstation is limited to 10 simultaneous connections and one remote access connection.

The software needed to set up a peer-to-peer network is built into most modern-day operating systems. Since 1995, Microsoft Corporation has introduced several desktop operating systems, each of which has peer-to-peer network capabilities. The most recent version of Windows provides the easiest and most reliable network connectivity: **Windows 2000 Professional**, **Windows XP Home Edition**, and **Windows XP Professional**. Peer-to-peer networks are also available in Windows 95, Windows 98, Windows NT Workstation 4, and Windows Me, although these versions are becoming less and less common in most organizations.

There are a lot of similarities between Windows 2000 Professional and the Windows XP versions. Both make use of an icon called My Network Places for viewing resources available on the network. By right-clicking on the My Network Places icon and selecting Properties, you can easily view your current configured network connections. Available in both versions of Windows is an auto-detection feature that can identify your network hardware and automatically set up the software and drivers that will work with your network card and on the most common networks.

The exercise on the next page will take you through the process of installing a network **adapter**, selecting a **protocol driver**, and adding file- and print-sharing capabilities. If you haven't added or configured Windows 2000 or XP before you will want to pay close attention to the order of the steps involved in configuring the network settings. Understanding the configuration is an important part of troubleshooting. So, even though your computer may have auto-configured your network settings for you, you still want to know how to configure your computer manually in the event that the computer experiences a network-related problem.

> **NOTE**
> If your copy of Windows 2000 Professional auto-detected your network card, then Client for Microsoft Networks and the TCP/IP protocol were automatically installed, along with the software drivers needed for the network card.

Network Architectures

Test It Out: Configuring Peer-to-Peer Networking

In Windows 2000 Professional, configuration begins with the Network control panel, which can be accessed through the Control Panel folder under Settings or through the Properties page of My Network Places. Before you begin this exercise, you will need to have a NIC installed. If you install one, Windows 2000 or XP should detect the card for you automatically.

1. Open the Network control panel: click on the Start Menu, select Settings, and click Control Panel. In the Control Panel folder, double-click the Network and Dial-Up Connections icon.

2. If you had a network card installed, the Local Area Connection icon will appear in the folder. Right-click on the icon and select Properties. The General tab displays the current protocol, network client, and network card adapter being used for the connection.

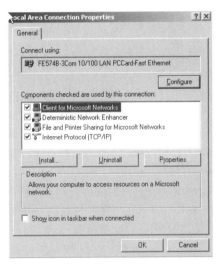

3. To add a new protocol or client, click the Install button. A list of options appears for installing a new client, service, or protocol. Select the protocol, and click Add.

Windows XP Home Edition
Micrsoft's latest version of its Windows desktop operating system which has been trimmed down for home users. Many of the features of XP Professional have been removed from XP Home Edition.

Windows XP Professional Edition
Like XP Home Edition, XP Professional is Microsoft's latest operating system for the business environment. Many more networking features are available in the Professional Edition.

adapter
A general term for a network interface card.

communications protocol
A set of rules that defines how communications will occur.

4. Select your protocol from the list provided and click Add again. You'll be redirected back to the Local Area Connection Properties Page.

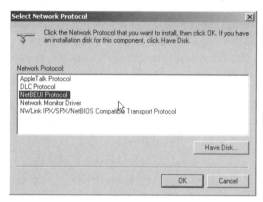

5. As discussed earlier, all of the necessary files, protocols, and drivers are installed by default when Windows 2000 detects the network card. That includes the File and Printer Sharing for Microsoft Networks.

6. You need to configure the computer name and the workgroup name to allow your computer to participate in peer-to-peer networking. To do so, right-mouse-click on My Computer on the desktop, and select Properties.

7. In the System Properties page, click the Network Identification tab. You can either select the Properties button to configure the identity manually or the Network ID button to use the Network Identification Wizard. In this exercise, you will use the Properties button.

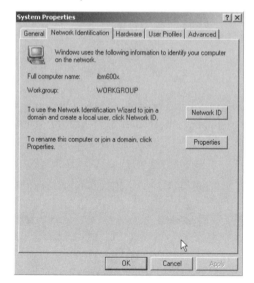

Network Architectures

8. You can change the name of your computer and the workgroup in the Identification Changes window. Enter the name that you would like to use and click OK. Click OK in the System Properties Window. Your computer will then prompt you to restart. You must click OK for changes to take effect.

Sharing Folders in Windows

The final step in configuring your peer-to-peer networking is configuring folders to be shared. Sharing a folder will enable others on the network to access the contents of those folders.

One of the great advantages of Windows 2000 over Windows 98 and Windows Me is that you can manage who has access to your shared folders and printers by using the Users and Passwords control panel. Unlike in Windows 98, in which you assigned a password to a resource, Windows 2000 allows you to configure unique usernames and passwords. That allows for more detailed control over file, folder, and printer security.

Test It Out: Creating Users

1. Open up the Control Panel folder, and double-click the Users and Passwords icon.

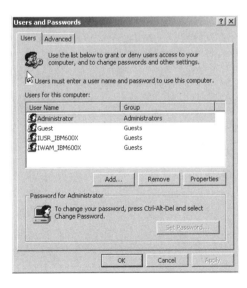

2. Click the Add button and fill out the user information. Click Next and enter the user's password. Click Next.

3. In the next screen, assign the user their level of access rights for the local computer. If they will be accessing only shared files and folders over the network, then select Standard User. They will be able to log on locally to your computer, but they will have limited access. If the user will need to log on locally and install and run some applications (such as the first time Office is run), then you should give them Administrator access. You should give out Administrator access only if it is a last resort. As Administrator, the user has full access to all parts of the computer, including changing the password for the Administrator account.

Test It Out: Adding a Shared Folder

If you were successful in the earlier exercise, "Configuring Peer-to-Peer Networking," then you should have your computer working on a network. In this exercise, we will be able to make good use of your earlier work by sharing a folder for other users to access on your peer-to-peer network.

1. To add a shared folder to your Windows 2000 Professional computer, navigate to the C:\ drive on your computer, and create a new folder. For this example, call the folder Users.

2. Right mouse click on the folder and select Properties. The folders properties window appears. Click on the Sharing tab.

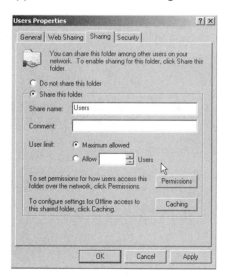

3. Click the Share this folder radio button. In the Share name field, which should now be white, enter the name of the shared folder that users will see if different from the name currently listed.

Network Architectures

4. Next, set who has access to the folder by clicking the Permissions button.

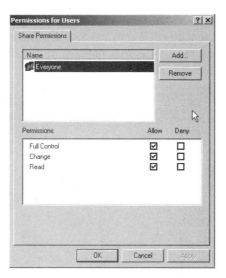

5. If you need to add a user, click the Add button and select the user from the list of available users.

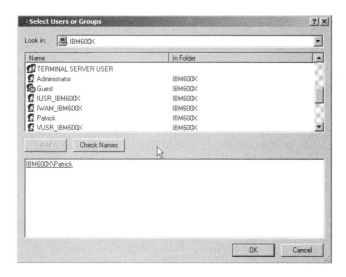

6. Once you have selected the user to add, click OK. In the Permissions for Users window, assign the new user the permissions that correspond to the level of access rights that you would like them to have. The highest level of access is Full Control, which gives them the right to add, change, and delete the folder and items in the folder.

53

Chapter 3

Accessing Shared Folders in Windows

Once file sharing is activated correctly, the icon for the shared folder changes. Before file sharing, the icon for the folder looked like a normal folder. When sharing is turned on, the folder icon is held up by a hand:

This means its contents are shared and are available to be accessed by other computers on the network.

Test It Out: Accessing Shared Folders in Windows

1. To access the shared folder over the network, you need to configure a second machine with the same workgroup name and protocol as your first computer. Log in to the second machine.

2. To access the computer with the shared folder, double-click the My Network Places icon on the Desktop. This will show the available workgroups and computers. You should see available workgroups, computers, and folders that are similar to the following folders:

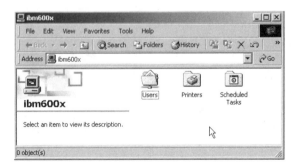

Network Architectures

3. Select the shared folder. In this example, the Users folder has been shared on the computer named IBM600X.
4. Open the shared folder to view its contents.

Configuring Macintosh Peer-to-Peer Networking

Apple Computer gets a lot of credit for being one of the first companies to create an operating system that made it easy to share files between computers. Apple's latest operating system is called OS X. On previous versions of the **Macintosh** computer, peer-to-peer networking was made possible with software called **AppleShare**. AppleShare is available in Mac OS 9.2.2 and earlier. But OS X, with its UNIX kernel and applications, does away with AppleShare. OS X takes a more robust UNIX approach to security by allowing you to customize the security settings to force users to log in. Once a user has logged in to the computer, their security settings will limit their access to applications, files, and changing settings. This is an important feature in schools and businesses that need to allow multiple users access to the same computer.

Macintosh
A type of computer that Apple Computer introduced in 1984 that was distinguished by its easy-to-use graphical user interface; now called the Mac OS.

AppleShare
The network operating system developed by Apple Computer that runs on a Macintosh server.

NOTE

On Macintosh computers, the term *user* refers to anyone to whom you want to assign a username and a password so they can access your computer. Users can then be organized into groups. Groups can be given security access to a folder or file just like users. If you grant a group access to a shared folder or file, then everyone who is a member of the group has the same level of access to the shared folder or file.

Test It Out: Configuring OS X File Sharing

To share folders on a Macintosh, do the following:

1. From the Doc (located at the bottom of your screen), click the System Preferences icon.Sharing icon under Internet.

2. Under Internet and Network, click the Sharing icon. Assuming you have administrator access to your Mac, click on the Start button for File Sharing. There is no need to turn on Web Sharing.

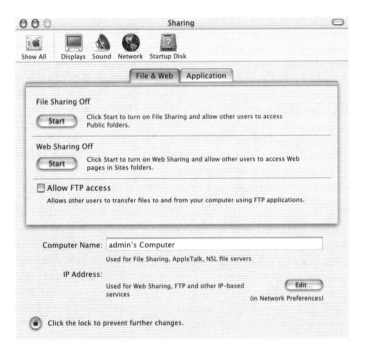

One of the interesting points about file sharing in OS X is that there is only one folder that is shared when file sharing is turned on. By default, the Public folder for your user account (there is one for each user who has an account on the local Mac) becomes available on your local network.

3. The next step is to add users using the Users application located in the System Preferences. Click on the New User button to add the user.

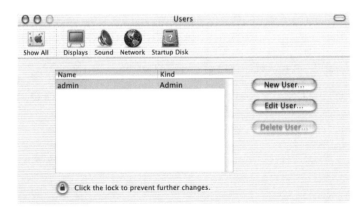

Network Architectures

4. Add the information for the new user. Click the Password tab to assign the user a password. If the user needs to have administrator privileges to the computer, then check the "Allow user to administer this computer" check box.

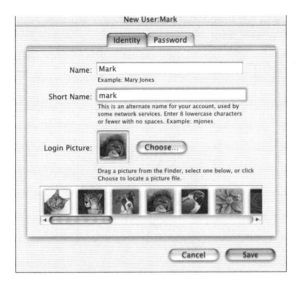

access rights
Security properties assigned to a network object or resource that define the level of access for a user. Common levels of access rights include read, write, change, and no access. Also known as *access permissions*.

5. Click Save to add the user to the list. The following prompt will appear:

6. If you want to force all users to have to log in to use the computer, uncheck the box labeled "Turn Automatic Login On." Otherwise, you can keep it off, and users will be able to boot directly to the desktop.

7. Now you can assign users **access rights** to the information in the shared folder. You can assign privileges to files and folders by selecting the item and pressing the Command and I keys simultaneously. The Get Info window will appear, which will give you direct access to the item's security privileges.

NOTE

In order to change the privileges on any item, you need to be the owner of the item. Usually, the administrator of the OS X computer has ownership of all items, although this may not be the case in a corporate environment.

57

Chapter 3

Test It Out: Accessing Shared Files on a Macintosh

The process for accessing another computer with a shared folder has changed dramatically in Mac OS X. The Chooser no longer exists. What does exist is a simple option in the menu bar named Go. The Go menu represents the fast way to navigate to folders, servers, and applications.

1. After you have finished configuring your Macintosh for file sharing, go to another workstation.

2. If you are not in the Finder, click the desktop. You should see the Go option in the menu bar. Click Go and select Connect to Server.

3. If you do not see the computer in the list, enter the name of the computer in the Address field. Click Connect to view the shared folders on that computer. Unlike in previous versions of the Mac OS, OS X does allow you to access shared folders on Windows-based computers. In the address field, type in smb://servername/sharedfolder/ to access the Windows computer.

Network Architectures

Clients and Servers

The client-server network architecture is a centralized model for data storage, security, running applications, and network administration. It is the most common networking architecture in use today. Many companies quickly outgrow peer-to-peer networks and have to find better networking solutions. Most end up installing a client-server network of some kind.

service
An application on a server that provides greater functionality for the end user.

Client-server networks, also referred to as *server-based* networks, have the following characteristics:

- They're based on a scalable model that can support small networks of five to 10 users, as well as large networks with thousands of users.
- They employ servers—specialized computers that provide services to the client workstations.
- They provide **services** such as printing, file saving, and applications.
- They allow a high level of security based on access permissions.
- They can be centrally managed by a network administrator or a team of network administrators.

The client-server model requires a special hardware and software implementation. The model takes advantage of the user's workstation, called the *client*, to distribute part of the work between itself and the server. One way that the client participates is by running a client version of an application locally. Then the server can take on the major task of storing large amounts of data and processing requests made by many clients. The server is usually a much more powerful computer that is capable of handling thousands of simultaneous requests.

In the following example, Sue uses the e-mail program running on her workstation to request new e-mail from the server. The server processes the request and sends any e-mail messages for Sue back to her workstation.

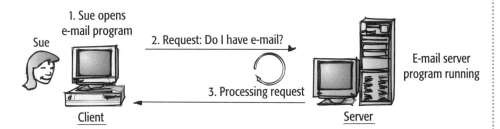

Chapter 3

Common Server Types

As a company grows, it is likely that the server it started with will become less and less capable of meeting its high demands. When performance begins to degrade, a company will add additional servers to perform specific tasks. It is not unusual for a large company to have two or more servers dedicated, for example, to respond to users' login requests.

print queue
A location on a server or workstation that stores print jobs that are waiting for a print device.

There are several types of servers. Some, such as file and e-mail servers, are more common than others, and are found at most companies. Server types include:

File servers   Unlike file sharing in a peer network, file servers offer users a central location to save files. Files stored on a file server are secure because they require a user to log in with a unique login name and a unique password.

Print servers   As networks grow, so do the printing demands of the users. Print servers help balance the load of printing by allowing users to print simultaneously to the server. The print server stores the print jobs in a **queue**—a temporary folder—until the printer is available.

Messaging servers   These servers answer requests for mail by clients or route mail messages to appropriate mail servers.

Application servers   When there is a heavy demand for an application, an application server improves performance and security by keeping the data and the application on the same computer. Many websites integrate two application servers: a web server and a database server.

> **Web servers**   Also known as HTTP servers, web servers give users access to information from any computer that has Internet access. The client computer runs an application called a *browser* that requests information from the web server.
>
> **Database servers**   A database is an application that stores records that contain information. Many companies keep large databases on servers so that clients can ask the server to process a database search or to generate a report.

TIP

For a workstation to be a client, it must be utilizing services from the server. Otherwise, it's just a workstation.

Network Architectures

Terminals and Hosts

Before there was a client-server model, there were terminals attached to a host computer. The host computer, called a **mainframe**, was responsible for all communication, storage, and processing. The mainframe was literally the brain of all of the terminals. In the terminal-host model, the **terminal**, often called a *dummy terminal*, ran the application from the mainframe. The terminal did not have its own hard drive or other means to store data. It was just a place to input keystrokes that would be displayed as output (text on the screen) and sent to the mainframe. All the actual work of storing the information or running the application was completed at the mainframe computer.

mainframe
Powerful computers that are much faster than most computers and are capable of processing and storing large amounts of data.

terminal
An input/output device that relies solely on a minicomputer or mainframe for processing, storing, and viewing data.

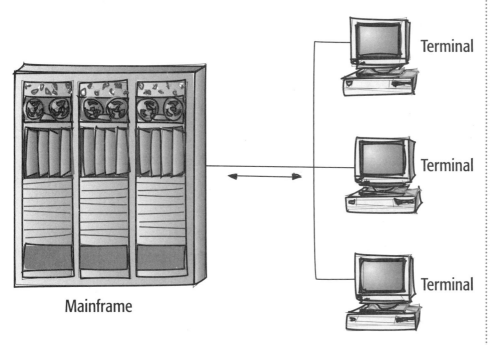

As the need for more powerful and larger systems grew, the terminal-host system became increasingly inadequate. At the same time, developments in computer technology made it affordable for most, if not all, users to have a computer on their desk that had the capability to store data, run applications, and process requests.

Selecting a Client-Server Network

Many companies have more than 10 users and, therefore, will probably be using a client-server network. Planning and selecting the right client-server technology is not easy. As you will see in the next sections, client-server networks can be implemented several different ways.

The basic criteria for selecting a client-server network are:

- Files need to be stored centrally.
- Security is important to protect sensitive and valuable data.
- Users will need access to the same application and data.
- A network administrator will be managing the server(s).
- There are more than 10 users.

NOTE

Make sure that you have considered the advantages and disadvantages before investing in a client-server network.

The advantages of client-server networks include:

- Data is stored centrally and can be easily backed up.
- A high level of security can be implemented at the server.
- Most powerful equipment can be shared.
- Server hardware and software are optimized for performance and reliability on client-server networks.
- Users are relieved of the burden of managing resources.
- The management of user accounts and resources is centralized.

The disadvantages of client-server networks include:

- Planning, design, and management are complicated.
- Managing servers requires dedicated staff.
- Server hardware and software are expensive.

Security on Client-Server Networks

Client-server networks have much better security than peer-to-peer networks. Although workstations are able to provide significant security locally on the workstation, it does not compare to the security available using servers. First, servers centralize resources on a network so that a single user account can be

Network Architectures

better managed. Managing the user accounts and the resources the users have access to is the responsibility of one (usually highly trusted) individual. Finally, servers offer stronger security applications such as encryption and sophisticated authentication systems. Let's revisit the earlier example comparing network security to keys and apartments:

In this example, the apartment tenants Sue, Julia, and Bethany live in the same apartment building. When they moved in, the apartment manager gave each person a separate key that was unique to their apartment. The apartment manager keeps track of who has what key. If someone loses a key, the apartment manager will reissue a duplicate only to the person who was originally assigned the key.

This analogy applies to coworkers Sue, Julia, and Bethany at work. They each have been assigned a password. The names they use to log onto the network, called *user* or *login* names, and their passwords differentiate one from the others. When the network administrator grants one of them permission to use a resource, they have been given a "key." The key applies only to that user.

Implementing Client-Server Models

Implementing server-based networks is not as simple as implementing peer-to-peer networks. There are many different ways to design server-based networks. Likewise, there are many different ways to configure hardware and software for servers. Selecting the right design depends on the size of your organization, the need, and the cost.

Chapter 3

> **NOTE**
> See Chapter 4 for information on server hardware and software.

Server-based networks come in three types, based on the following criteria:

- Number of users
- Application needs
- Geographical location

Single-Server Networks

When a company outgrows its peer-to-peer network, it usually adds a server and converts the network into a client-server network. Single-server networks are typically employed by companies with 10 to 50 users, though even small companies may opt for more than one server. A small business might add a second server for special-purpose applications or to make their network more reliable—a single-server strategy is risky because it increases the likelihood of data loss in the event that the server fails.

Single-server networks should be used only if just a few services will be running on the server. A simple single-server network is shown here:

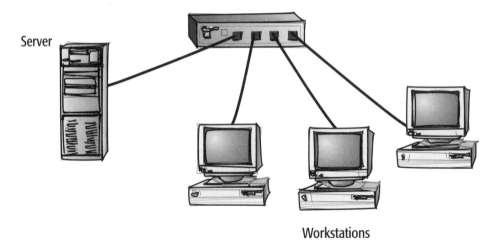

Workstations

Multiserver Networks

Because servers can provide a variety of functions, it is not uncommon to have several servers working on one network, each providing different services. This type of network, called a *multiserver network,* is typical of networks with 50 to

Network Architectures

500 users. Separating services across multiple servers improves performance and reliability. Each server can be optimized to run a service. If one service fails, other servers continue to function normally and can even take up the slack of the malfunctioning server.

A multiserver network is shown as follows; note that it is more complex than the single-server network.

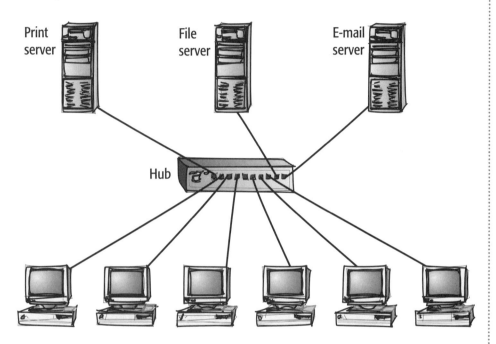

Enterprise Networks

As companies grow, so do their networks. Today, companies of all sizes are using networks so employees can communicate and work more efficiently. In some companies, a day's work may involve collaborating with coworkers located in offices around the world. These companies require networks that will support thousands of users who need to access information across the company. These large networks are called *enterprise networks*.

Enterprise networks can be enormous, with thousands of users and possibly hundreds of servers. Each office location may look like a single- or multiserver network, except that each location will be connected to the rest of the corporate wide area network.

Chapter 3

Client-Server Applications in the Enterprise Network

It is common today for a company to connect several local area networks, or LANs, that are located far apart from one another, to create a wide area network, or WAN. Wide area networking has become a testing ground for new client-server applications, which are capable of communicating between distant servers. This extends the client-server model across many LANs, involving several servers to fulfill a user request.

To get a better idea of what is happening with client-server models over WAN links, let's revisit our previous e-mail example. One user, Sue, was requesting a process from just one server on her company LAN. Now, Sue wants some information from another server in an office across the country. The servers on each of the LANs in her company's wide area network work together to fulfill Julia's client request for sending her e-mail message.

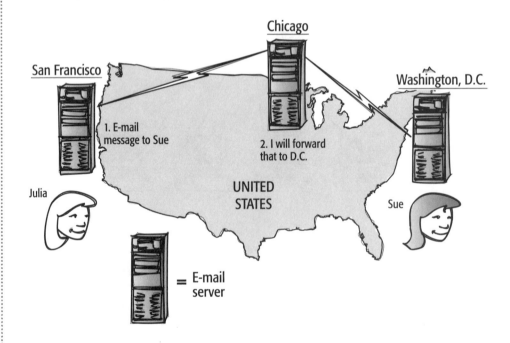

Network Architectures

Hybrid Networks

Hybrid networks incorporate the best features of workgroups in peer-to-peer networks with the performance, security, and reliability of server-based networks. Hybrid networks still provide all of the centralized services of servers, but they also allow users to share and manage their own resources within the workgroup. After a user in a workgroup logs in to the network, they don't have to have any other interactions with the server while they access shared files in the workgroup.

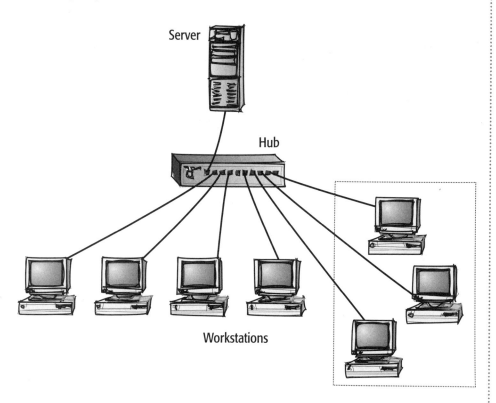

The advantages of hybrid networks include:

- Client-server applications are still centrally located and managed.
- Users can assign local access to resources on their computers.
- Workgroups can manage resources without requiring assistance from the network administrator.

The disadvantages of hybrid networks include:

- Network access can become burdensome for the users.
- Users may need to remember multiple passwords.
- Files can be duplicated and changes overwritten between the computer with the shared folder and the server.
- Files saved on the workstation are not backed up.

NOTE

In large corporate networks, users often do not have the ability to share files. The information technology department will typically provide users with the ability to use their workstation and nothing more. The user won't be able to make changes to their workstation. The reason for this tight security is to prevent the introduction of illegal applications and viruses, and to simplify workstation support.

Review Questions

1. What are three examples of network architectures?

2. Name three characteristics of peer-to-peer networks.

3. Describe security on a peer-to-peer network.

4. Define the term *server*.

5. List the three hardware components needed to connect computers together to make a peer-to-peer network.

6. Where should you begin network configuration in Windows 2000?

7. How do you allow folders and printers to be shared in Windows 2000?

8. What are three of the disadvantages of peer-to-peer networks?

9. What are two of the advantages of client-server networks?

10. Describe the differences between a file server and a messaging server.

11. What are two of the factors used to determine if a single-server, multiserver, or enterprise network is needed?

Terms to Know
- 100BaseT
- access rights
- adapter
- AppleShare
- client-server networks
- hybrid
- ISA
- Macintosh
- mainframe
- network administrator
- network architecture
- passwords
- PCI
- peer-to-peer networks
- ports
- protocol driver
- queue
- security
- servers
- services
- share-level security
- terminal
- Windows 2000 Professional
- Windows XP Home Edition
- Windows 98
- Windows NT Workstation 4
- workgroup
- workstations
- wrist strap

Chapter 4

Network Services and Software

In Chapter 3, we introduced the term *service* to describe a resource that is made available to users on the network. The term comes from the idea that the server computer provides a service to the workstations. The service may be printing, accessing files, or running an application.

The concept behind the use of a server is similar to the role of a waitperson in a restaurant: The waitperson is there to serve you. You choose an item from the menu and order it from the waitperson; when your dish is ready, they serve it to you. You may have to wait a long time for your food if the waitperson is slow or if the restaurant is busy. Also, the waitperson can only serve you what is on the menu.

You can evaluate the performance of a network server the same way that you rate the service that you receive from a waitperson in a restaurant. A server needs to fulfill requests in a timely manner. If the server takes a long time to respond to your order, you will become frustrated. It may be critical that the server responds quickly to time-sensitive tasks such as a database search or opening a multimedia file.

In addition to explaining servers and the services they offer, this chapter will cover the following related topics:

   Types of network operating systems

   Configuring network clients

   Testing the connections

Chapter 4

Types of Network Services

A server can provide a variety of network services. Some services are never seen by the user but influence the way the server functions. Others affect how users carry out tasks over the network; we talked about several of these services in Chapter 3:

- File services
- Print services
- Application services including databases and web servers
- Messaging services in the form of e-mail and news

There are many other services that are available with server software. Some of them are listed below:

Dynamic Host Configuration Protocol (DHCP) Services   This service dynamically assigns an IP address to workstations that request to communicate on an IP network. DHCP largely eliminates the need for manual entry and avoids duplicate IP address conflicts. DHCP is often provided by DSL routers and cable modems.

> **NOTE**
> DSL and Cable Internet Access are described in detail in Chapter 14.

> **NOTE**
> Manually entering addresses is known as *static addressing*. With static addressing, the network administrator manually configures every device that requires an IP address. Static addressing is usually fraught with human error.

File Transfer Protocol (FTP) Services   Although using a web browser for file downloads has made FTP less common, it is still used as a method for sending and retrieving files from a server using the TCP/IP protocol. The FTP server, also called the **FTP daemon** ("dee-mun") on UNIX systems, provides the service. The FTP client software for downloading files is useless without the FTP server.

Domain Name System (DNS) Services   The DNS service is a database that maps the IP address of a web server or other device to the "real" name that we use on the World Wide Web. When you request the website www.sybex.com, the request is sent to the DNS server first. The DNS server sends back the IP address so that you can contact the website.

Network Services and Software

Virtual Private Network (VPN) Services   The combination of hardware and software that allows users and network administrators to access the internal network of the organization from any connection to the Internet. Unlike remote access solutions that use modems and phone lines, VPNs help organizations reduce costs by lowering phone charges. Some organizations do not use VPNs because the data, although encrypted, still must travel over the highly insecure network of the Internet.

> **NOTE**
> Remote access servers still exist in many organizations for dial-up users who require access from locations such as hotels or where the highest level of security is required.

Chapter 4

Choosing a Server

Usually, when we talk about a server, what we are referring to is the software that runs on a computer called a *server*, not the computer itself. This software is the **network operating system (NOS)**. There are many different kinds of network operating systems available. Although the underlying technology for each network operating system may be different, the basic services each provides are generally the same.

network operating system (NOS)
Software designed specifically for use on a network server to provide multiple services and a high level of performance.

What to Look for in an NOS

Although different NOS software packages may provide similar services, they may not all offer the same level of performance, flexibility, security, or scalability. It is important to look for features that will meet the needs of your small business, school, or company while addressing the requirements of increased network demands.

multitasking
The processor's capability to perform several tasks simultaneously.

Multitasking

Servers need to respond to many requests quickly in order to help people stay productive and not become frustrated. Server software should be able to process requests and complete other tasks at the same time. The ability to perform several tasks simultaneously is called **multitasking**. Some pieces of server software perform multitasking better than others using a process called **pre-emptive multitasking**.

pre-emptive multitasking
The capability of some NOS software, such as Windows NT and UNIX, to manage tasks so each application gets equal access to the processor.

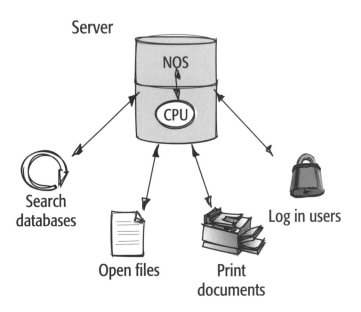

Network Services and Software

Multiprocessing

Even if your server software can multitask, there may come a point when the demand on the server is too high. When the **central processing unit (CPU)** is overutilized by too many requests, the server becomes sluggish. If you cannot upgrade to a faster **processor**, you need to be sure your network operating system supports **multiprocessing**. Multiprocessing improves performance by allowing you to add more CPUs to the same server computer. When a server has two or more CPUs, the NOS manages the tasks evenly to each CPU. The demand for multiprocessing servers is very high in large companies. Don't be fooled, though; most small businesses and schools don't need multiprocessing. So make sure you understand your performance needs before investing in this costly feature. In addition, if you are going to have multiple processors you will need to make sure that your applications (including the NOS) have been designed to take advantage of the multiple processors.

central processing unit (CPU)
A microprocessor chip, the CPU is the brain of the computer; it uses mathematical functions to perform calculations and complete tasks.

processor
Another term for CPU.

multiprocessing
Multiple processors within a device working on the same task.

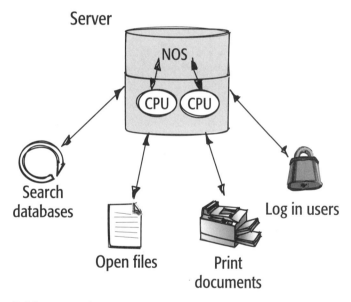

Finally, a reliable network operating system is stable. Stability in a network operating system is defined by how well the server performs under heavy load and in spite of any software errors. In a stable system, if an application running on the server unexpectedly quits or stops functioning, the server software will continue to operate normally. The errant application will be terminated and restarted without shutting down the server.

NOTE

Learn more about different server hardware in Chapter 13.

Chapter 4

Server Hardware

In addition to choosing a network operating system, you'll need to consider the type of server computer that will best meet your needs. Servers should be selected based on the following criteria:

Reliability   The server is capable of providing continuous service in the event of a critical failure or increased demand.

Scalability   As demand and needs grow, the server can be upgraded to accommodate growth.

Fault tolerance   The server minimizes the possibility of data loss by using a variety of hardware and software to recover from faults or failures.

The more reliable, scalable, and fault-tolerant a server is, the more it will cost. Servers that have been designed with all three criteria in mind can be quite expensive. But the cost of the server is usually miniscule in comparison with the cost to re-create your data.

When selecting your hardware, you will need to include a tape backup drive and decide on an uninterruptible power supply (UPS). Depending on your needs, you may need to add other options, such as multiprocessing or a RAID system. Adding these features to your server will increase its **fault tolerance**, thereby protecting user data and avoiding as much downtime as possible:

Tape backup drive   A tape backup drive uses software and hardware to back up data on the server. The files on the server are copied to a magnetic tape daily. The tapes should then be stored in a safe location where they won't be damaged. This is a slow but inexpensive way to store large amounts of data. Larger companies send tapes off-site for storage and keep another set in their own safe.

Multiprocessing   A fast processor is important, but servers need to be able to add a second processor (or more) if performance demand increases dramatically.

RAID (Redundant Array of Inexpensive Disks)   Servers usually have two or more **hard disks** for storing data. A RAID configuration includes hardware and software that protect data in the event one of the hard disks stops working. There are five RAID levels, with RAID level 5 offering the best insurance against drive failure. RAID level 5 also improves the speed of saving data to the hard disks by allowing all disks to write information simultaneously.

Uninterruptible power supply (UPS)   The UPS is a battery backup system. In cases when power is lost, the UPS provides temporary power so that the server can be shut down properly, avoiding any loss of data.

---

fault tolerance
Designing a system that will continue to perform operations, even if interruptions to power or hardware failure occur.

hard disk
A large-capacity magnetic storage device with read and write capabilities.

Network Services and Software

Types of Network Operating Systems

As we discussed earlier, most network operating systems provide the same basic functions. Even so, it is important to understand how the different brands and versions perform. Where one type of server software may offer ease of installation, another may be more stable. Selecting the right server software requires knowing the needs of your organization.

Windows 2000 Server

Microsoft Windows 2000 Server is the foundation for Microsoft's latest server technology. Like most software upgrades, Windows 2000 Server introduced a number of enhancements and features. Taking some of the best features of Windows 98, such as zPlug and Play, and the well-known functions of Windows NT 4.0 was just one part of Windows 2000. Windows 2000 Server represented a major departure from previous versions. First, Windows 2000 Server comes in three models: Standard, Advanced, and Data Center. Each is designed to meet the particular demands of organizations. The following chart represents just a few of the differences between the three server products.

	Windows 2000 Server	Advanced Server	Datacenter
Max CPU	4	8	32
Max Memory	4GB	8GB	32GB
Failover Clustering	No	Yes	Yes
Load Balancing	No	Yes	Yes

The introduction of Windows 2000 Server also marked the debut of Active Directory. Active Directory is Microsoft's vision of providing directory services similar to that found in UNIX versions and Novell NetWare. Active Directory gives IT administrators a single view of their organization's technology resources and how they are allocated. Active Directory greatly simplifies the management of users and resources for a particular department or location.

For large organizations, Active Directory can be replicated between multiple domain controllers. No longer is an organization dependent on the reliability of primary domain controller, as was required in Windows NT 4.0 Server.

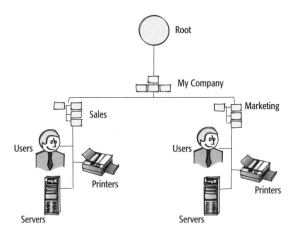

Should You Choose Windows 2000?

Windows 2000 has made server administration even more reliable and simpler to manage. Although Active Directory is considered to be more complicated, it provides an expandable platform for integrating other products. And knowing that there is a growth path that is competitive with UNIX counterparts should give administrators confidence in their decision.

The advantages of using Windows 2000 Server are:

- It's easier to manage with improved management tools.
- Fewer reboots are required when changing the configuration.
- It provides improved hardware support with Plug-and-Play capability.
- It has improved stability and uptime.
- There is support to grow from a single processor server to large-scale servers.

The disadvantages of choosing Windows 2000 Server are:

- Proprietary technologies may limit flexibility.
- There are high licensing costs for companies.
- It is not as stable or as fast as UNIX versions.
- It has high CPU, disk, and memory requirements.

Network Services and Software

The minimum system requirements for Windows 2000 Standard Server are the following:

	Manufacturer	Recommended
CPU	Pentium 133	PIII 600
Memory	256MB	512MB
Hard Disk	1GB	10GB

UNIX

Just speaking the word UNIX, pronounced "you-nix," can strike fear into the heart of the average computer user. That is assuming they have even heard of UNIX. In some ways, **UNIX** is an operating system just like Windows or the Mac OS. It runs on a computer and allows the user to run applications, store information, input data, and output information to a printer. That is where the similarities end. UNIX is fast, reliable, and powerful. UNIX has long been used for engineering applications and scientific research. It has more recently found a role as the premier platform Internet application, powering the majority of websites on the Internet.

It is misleading to use the term *UNIX* to describe all UNIX software. UNIX was originally developed in the 1960s by Bell Laboratories at AT&T. Since then, many different variants have evolved. The most notable is Berkeley Software Distribution UNIX (**BSD UNIX** or Free BSD). Students at the University of California, Berkeley, developed BSD UNIX using the original software from AT&T. They reprogrammed the entire operating system, creating their own OS. Then they gave it away for free. Eventually, several companies developed their own variants of UNIX based on BSD. The list of UNIX flavors includes versions developed by the biggest computer software companies in the world: Sun Microsystems's Solaris, Hewlett-Packard's HP-UX, IBM's AIX, and (most recently) Apple's OS X server.

NOTE

There are many variations of UNIX. Some are designed to run only on special processors, whereas others will operate on several different types of processors. Determine your needs carefully before selecting a particular flavor of UNIX. Although there are some free types of UNIX available—such as Free BSD and Linux—you may find that a version developed, sold, and supported by an established software company is more reliable for a particular task.

UNIX
An operating system that supports multiple users, multitasking, and (in many cases) multiple processors. UNIX was created by AT&T in 1969. Today, there are several variations of UNIX.

BSD UNIX
A version of UNIX developed at the University of California, Berkeley. BSD UNIX is available for free from the Berkeley Software Distribution user group.

Whatever happened to AT&T? AT&T continued developing its version of UNIX: System V. The latest version, System V Release 4 (SVR4), was jointly developed by AT&T and Sun Microsystems.

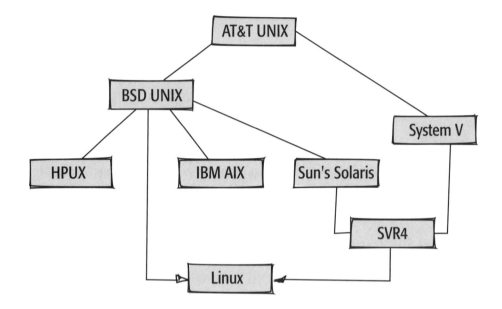

Should You Choose UNIX?

When you are selecting server software, UNIX should be a consideration if you have a network administrator who is experienced with UNIX.

The advantages of using UNIX are:

- It's a very stable operating system.
- It can function as a workstation or a server.
- It's fast—very fast.
- It includes hundreds, possibly thousands, of built-in tools and applications, including programming tools.

The disadvantages of using UNIX are:

- It's complex and uses archaic commands—learning UNIX is often compared to learning a foreign language.
- The software can be very expensive (although BSD UNIX is free).
- Some companies sell versions of UNIX that will run only on their hardware.
- It includes hundreds, possibly thousands, of tools, applications, and commands.

Network Services and Software

Linux

How popular is Linux? Linux has spawned dozens of software companies, specialized applications, and even Linux-only hardware companies. Some may say that Linux development was responsible for bringing Open Source software to the attention of the technology world. The use of Open Source code such as FreeBSD has even made it into Apple's Mac OS X. Needless to say, Linux and all of its other Open Source relatives are here to stay.

Linux is a version of UNIX that was originally developed by Linus Torvalds at the University of Helsinki. From the start, Linus made the software available for free to other UNIX programmers around the world, following the concept of the Open Source community. The idea is simple, but extremely powerful. Linus makes his code available for free to anyone who wants it, but in exchange, any changes that are made must be made available to the community. No one has the legal right to claim the licensing as his own. A smart idea for harnessing the mental power and time of developers.

The Linux source code can be downloaded from the Internet for free. The only difficulty is that you need to know how to compile the code using a compiler application. Although not an impossible task, compiling the source code for Linux is not as easy as the point-and-click installations of Windows-based applications.

Fortunately, several companies have made installing Linux straightforward and easy. They have even added great features to the software and some applications of their own. A few of those companies are Caldera Systems, Mandrake, and Red Hat Software; and they provide distributions of the Linux operating system with easy-to-use installations and, most recently, Plug-and-Play hardware installation. These companies charge a minimal amount for the software, around $50.00, to cover the costs of distribution and packaging. Almost all software companies that distribute Linux provide several different types of interfaces that look similar to Windows, such as XFree86 and Gnome. If you are new to Linux, then pick up a copy of one of the more popular versions. It will make installation a breeze.

It's not surprising why Linux is popular:

- Linux runs on computers using Intel 386 or faster processors.
- New applications are being developed for Linux all the time; most are free.
- Linux itself is free.

Linux
Linux is a variant of UNIX that was originally created by Linus Torvalds at the University of Helsinki, but now includes some of BSD UNIX and SVR4.

> **NOTE**
>
> Selecting a particular release of Linux can be confusing. Red Hat Linux is very popular, but other excellent choices abound—including Caldera's OpenLinux, SuSE Linux, Slackware's Linux distribution, Mandrake, and many more. Of course, if you are feeling particularly ambitious, and you have a lot of time on your hands, you can always download the actual software code itself at http://www.linux.org.

Should You Choose Linux?

If you are considering using UNIX, you are probably considering Linux. Although there are many reasons to choose Linux, consider the less-polished edges of Linux as well.

The advantages of using Linux are:

- It's as easy to install as Windows XP, and takes less time.
- Like UNIX, it's fast and reliable.
- It runs on PCs, PowerPCs, and SPARC stations.
- It's free online or for a minimal cost for a packaged product with CD.
- Hundreds of free software applications are available for Linux.
- Because the source code is available, bugs and security threats are fixed much faster than in any other NOS.

The disadvantages of using Linux are:

- It's still UNIX, which means there are hundreds of commands and applications to learn.
- Although Linux makes UNIX administration easier, it still requires a UNIX administrator.
- Only a few companies offer dedicated tech support for Linux for a fee. If you are relying on Linux as your primary server, then budget for phone support and an experienced consultant.
- Linux is currently not considered a replacement for critical applications that require UNIX.

Network Services and Software

The minimum system requirements for Red Hat Linux 7.0 are shown as follows:

	Manufacturer	Recommended
CPU	Intel 386	Intel 75
Memory	16MB	32MB
Hard Disk	125MB	1GB

Novell NetWare

In the mid-1980s, a network software company called Novell, Inc. was beginning to dominate the server-based network market with its **NetWare** NOS software. By the late 1980s, NetWare version 3.11 was the dominant server software for file and print sharing. Novell had managed to corner the growing market of client-server networks.

In the early 1990s, Novell lost its lead to Windows NT Server. Although Novell was deeply entrenched in the market, the NetWare product was designed for small networks that required only one server. As networks grew, there was no way to easily copy user information and passwords to other NetWare 3 servers. If a NetWare administrator needed to add a second or third server, they were forced to re-enter all the user information.

Novell introduced NetWare 4.11 in 1995; this release included improvements that made implementing NetWare in an enterprise network more manageable. It offered a significant improvement in its support for the TCP/IP protocol. (Earlier versions only supported Novell's proprietary protocol, IPX. Although IPX was heavily used on LANs, access to the Internet required IP.) Novell also introduced its new directory service, **Novell Directory Services (NDS)**. Just three short years later, NetWare 5 was introduced, with native support for TCP/IP and a number of the most popular Internet applications.

In 2001, Novell released NetWare 6. Building on NetWare 5, NetWare 6 comes with full support for TCP/IP improved support and functionality in its Internet applications. In addition, Novell's Directory Service now supports open standards for directory service protocols such as Lightweight Directory Access Protocol (LDAP), which allows it to be integrated in to a mixed NetWare, Windows, or UNIX environment.

NetWare
A network operating system created by Novell.

Novell Directory Services (NDS)
A set of software services created by Novell to provide access to a directory of information about network entities. It allows an administrator to organize and manage users, servers, and other devices on the network.

Novell Directory Services (NDS)

Like Active Directory in Windows 2000, NDS organizes a network into a structure called the *directory tree*. Under NDS, a tree grows as objects are added to it. Each object represents a user or device on the network—including a server, workstation, printer, or other network device.

Network administrators like NDS because it allows them to manage and view multiple NetWare, NT, and UNIX servers; as well as other devices, from a single location. They no longer need to use multiple applications to monitor performance or to manage the server. In an effort to maintain its presence on today's networks, Novell has made its Novell Directory Service available as a separate product that can run on other operating systems.

Should You Choose NetWare?

As with any product, Novell NetWare is not without its faults, but it does provide a reliable alternative to Windows 2000 Server.

The advantages of using NetWare 6 are:

- It boasts an extremely stable network operating system.
- Its open standards directory service, NDS, centralizes management of all network resources.
- If offers management support for NetWare, NT, and UNIX servers.
- It provides a Java-based management utility called ConsoleOne that takes advantage of NDS (and is free!).

The disadvantages of using NetWare 6 are:

- Like UNIX, administering NetWare 6 from the server requires that the administrator learn many commands.
- NDS can be more difficult to implement in large networks than non-directory service based systems.
- NetWare 6 is not as widely implemented as Windows 2000 or UNIX.

The minimum system requirements for NetWare 6 are shown as follows:

	Manufacturer	**Recommended**
CPU	Pentium II	Pentium II processor
Memory	128MB	512MB
Hard Disk	1GB	10GB

Network Services and Software

Mac OS X Server

If you have never considered Apple as a viable choice for server software, it's time for you to think differently. Not only is Mac OS X Server an option, it may be one of the best options for organizations of any size. Why? Well, it is UNIX with a Mac face. The Windows zealots should pause before making any comments. There is something very different under the hood of Apple's new NOS.

Mac OS X Server, released in March of 1999, is unlike any other server software Apple has ever released, and possibly unlike any other server software on the market. The core of Mac OS X Server is based on two versions of UNIX: BSD UNIX and Mach, a variant of BSD UNIX developed as part of the Mach Project at Carnegie Mellon University. Apple has also made a good deal of the Mac OS X core software, called *source code*, available for free. The hope is that software programmers will be inclined to further develop and improve the server software in the same way BSD UNIX and Linux have been developed.

Describing Mac OS X server as just another variant of UNIX doesn't even begin to touch the service of the capabilities and usability of OS X. Like any version of UNIX, OS X server can run almost any version of UNIX software that runs on the Mach kernel and FreeBSD.

Included in Mac OS X server are several standards-based applications, such as the following:

- Standards-based DNS, FTP, and DHCP
- Apache Web server
- Mail Server with WebMail
- Network File Sharing (NFS) for exporting shared drives on other computers
- Directory services such as Windows, Novell, and UNIX, using Netinfo and LDAP

In addition, Mac OS X server sports these additional features:

- A user-friendly management tool for all server applications, called Server Admin
- Seamless client support for all versions of Windows
- Quicktime streaming server for on-demand video for the Web
- Command-line access to all configuration features just like UNIX (available, but not required)

Probably the most overlooked feature of Mac OS X Server is that you have a robust and scalable NOS that is tightly integrated and tested on Apple hardware. With support for symmetric multiprocessing and pre-emptive multitasking using G4 processor technology, Mac OS X Server is fast and it works.

The advantages of using Mac OS X Server are:

- It has improved server security using easy-to-use applications.
- It's based on the reliable BSD UNIX and Mach software.
- It includes many standard Internet server applications, including Apache Web server.
- It supports centralized management of Macintosh computers.
- Its Open Directory architecture can be used for integrating user authentication and folders with UNIX, Novell, and Windows 2000 Active Directory.
- A rack mount server, called Xserve, is now available and is only 1.75 inches high.

The disadvantages of using Mac OS X Server are:

- With a small user base, it's less proven than other NOSs.
- It does not support existing Macintosh Server applications.
- The performance is limited to the hardware options available from Apple.

Apple's minimum system requirements for Mac OS X Server are shown as follows:

	Manufacturer	**Recommended**
CPU	G3 (iMac)	G4
Memory	128MB	256MB
Hard Disk	4GB	20GB

Client Network Configuration

If you've already configured a workstation to participate in a peer network, then configuring a client workstation to communicate with a server will be easy. The only major change will be the installation and configuration of the TCP/IP protocol. As you learned in Chapter 1, the TCP/IP protocol allows computers to communicate over the Internet and is the communication protocol used by UNIX computers.

Configuring Windows XP Clients

Whenever you configure network information on any computer, make sure to verify that the information is accurate. If you are working on a Windows XP computer that already has the TCP/IP protocol installed, you can go ahead and change your TCP/IP settings without having to reboot. In Windows XP Professional, you can also assign multiple IP addresses to the same network card. This may be useful on a computer that has settings you want to keep, and you are only testing your new TCP/IP settings.

Windows Networks

Users who want to access a server on a Windows 2000 domain need to log in to the domain first. When users are prompted to log in—a process called **authentication**—the window will specify only the domain, not the server. This may seem odd at first, but it makes sense. You will recall that Windows 2000 uses domain controllers that are identified in Active Directory. The Active Directory database is located on all domain controllers. In the event that one of the domain controller servers fail, the next-closest domain controller can authenticate you. Authentication would not work well if the users had to rely on a single server for authentication and access to resources, although this is often the case in smaller organizations.

authentication
The process of a server validating the username and password submitted by a user during login. If the authentication is successful, the server returns the user's security rights; if unsuccessful, the user is denied access.

Test It Out: Connecting to a Windows 2000 Network

1. On a Windows XP computer, navigate from the Start to the Control Panel icon. From there, you will need to click on Network and Internet Connections and then on Network Connections, which is located under "or pick a Control Panel icon". Client for Microsoft Networks should already be installed and listed. Double-click it.

   In order to connect to the domain, you will need to have an administrative account. If you do not have an administrative account for your Windows 2000 Server, or if you are not sure, you will need to contact your network administrator.

2. Select the Log on to Windows NT Domain radio button.

3. In the Windows NT Domain field, type the name of the domain and click OK. If you are able to add your computer to the domain, you will be prompted to reboot your computer.

Novell NetWare Networks

If you have to connect to a NetWare server, you need to add the Client for NetWare Networks.

Test It Out: Connecting to a NetWare Network

1. Open the Network Connections control panel, right-mouse-click the Local Area Connection icon and select Properties. Click the Install button under the list of installed network components. Double-click Client.

2. Select Client for NetWare Networks and then click OK.

3. After you reboot your computer, you will be prompted to enter the necessary NetWare server information. If you don't know the name of your preferred tree or name context, contact your network administrator.

Network Services and Software

TIP

Novell recommends that you install the Novell Client software that comes with Novell NetWare. The Microsoft software for connecting to a NetWare server will work, but it does not perform nearly as well as the Novell software.

Configuring Internet Addresses

The only part of the TCP/IP protocol that you need to configure is the IP address, subnet mask, and **default gateway**. The IP address for every device connected to the Internet is a unique number. Much like your home street address, the IP address of your computer distinguishes it from all other computers in the world. If you are on a LAN, the network administrator assigns an IP address to your computer. If you use DSL, cable, or a dial-up modem to connect to the Internet, your **Internet service provider (ISP)** assigns your IP address.

TCP/IP on Windows XP Computers

If you have a network card (most newer computers do now), then XP automatically installed the TCP/IP protocol. Fortunately, that is one less thing you have to worry about. Windows XP does not install any other protocols by default. This may be a concern if you are planning on accessing the Windows XP computer using computers running Windows 95 or 98. In order for Windows 95 and 98 to connect to shared files, folders, or printers on the Windows XP computer you will need to install the **NetBEUI** protocol.

NOTE

The configuration of TCP/IP for Windows XP begins with assigning the IP address. If your company is using the dynamic IP address assignment that is provided by a server, you won't need to do anything. Dynamic IP address assignment is the preferred method for setting the IP addresses of computers in companies, and for DSL and cable Internet users. If you have to manually assign your IP address, you will need to complete the following steps. The IP address and the subnet mask distinguish your computer and network from all others on the Internet. Ask your network administrator for this information.

Internet service provider (ISP)
A company that offers (to businesses and the public) access to the Internet by selling services such as dial-up accounts that use modems.

default gateway
A device, usually a router, which connects two or more different networks together.

NetBEUI (NetBIOS Extended User Interface)
Pronounced "net-booee," NetBEUI is a Microsoft protocol that was originally designed for Microsoft LAN Server software. NetBEUI is considered a fast protocol, but it cannot be routed between networks.

89

Chapter 4

Test It Out: Configuring TCP/IP on a Windows XP Computer

1. Open the Network Connections control panel.
2. Right mouse click the Local Area Connection icon and select Properties.
3. Double-click the "Internet Protocol (TCP/IP)" component.
4. By default, your TCP/IP settings are set to obtain TCP/IP information dynamically. To configure static entries, click the radio button "Use the following IP Address".
5. In the IP Address field, enter the IP address assigned to you by your network administrator. You can use 192.168.1.3 for practice.
6. In the Subnet Mask field, enter the subnet mask. If you are using the sample IP address from the previous step; the subnet mask is 255.255.255.0.
7. In the Default Gateway field, enter the IP address used to reach the Internet. This often refers to the router address. The first IP address is commonly used by administrators. Enter 192.168.1.1 for practice.

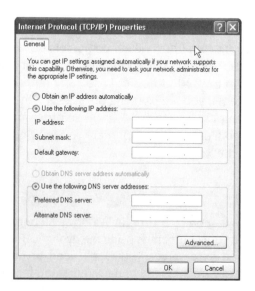

8. Leave the Internet Protocol Properties window open for now. You will need to configure the DNS settings, which are described in the next section.

Network Services and Software

> **NOTE**
>
> The IP address 192.168.1.3 is called a *private address space*. There are more than 64,000 usable numbers in this set, from 192.168.0.1 through 192.168.255.254. These numbers are called *private* because they cannot be used on the Internet. They are for internal use by companies; therefore, they can be used over and over again. In addition to the 192.168 block of addresses, 10.0.0.0 and the block of addresses from 172.16.0.0 to 172.31.0.0 are all private addresses that can be used for internal networks.

domain
In Microsoft terminology, *domain* refers to the domain model for organizing NT servers within a company.

The Domain Name System (DNS)

When configuring DNS for your computer, you need to fill in several pieces of information: The *host name* is the name of your computer; it is best to use the computer name that you gave in the Identification tab of the Network control panel. The **domain** is the Internet domain that represents your company (mycompany.com). The term *DNS server* means the IP address of a DNS server, whereas the *domain suffix* is the "real" name of a DNS server.

Test It Out: Configuring DNS

1. Continuing from where we left off in "Test It Out: Configuring TCP/IP on a Windows XP Computer," click the radio button "Use the following DNS server addresses" if it has not been already selected.

2. Enter the IP address of the preferred DNS server, which is more commonly known as the primary DNS server. If you don't have a real IP address available from your network administrator, you can use 192.168.1.10.

3. In the Alternate DNS Server field, enter a second DNS server. Use the IP address 192.168.1.11. The second DNS server provides a backup in the event that the primary server is not available.

4. Click OK to apply your new TCP/IP settings.

Testing the TCP/IP Configuration

After completing the TCP/IP configuration, you can test your configuration using a few simple commands in Windows XP. One of the commands, ipconfig, allows you to view and confirm your IP configuration. To use this tool, go to the Start menu and choose Run. In the Open field, type **cmd** and click OK. A Command window will appear. At the Command prompt, enter the command **ipconfig** and press OK. You should see output similar to this graphic:

Ping
An application used to test whether a connection exists between two remote devices on the Internet. Requires the TCP/IP protocol.

You also can check the connection between your computer and another by using two utilities: Ping and Trace Route. **Ping** is an IP utility that tests the connection between two computers. The Ping utility sends a message to the IP address that you specify. If the recipient receives the ping, it sends a reply back to you. To ping a computer, use the same Command window as you did with ipconfig, type **ping** followed by a space, and then enter the IP address of the computer you want to test on your network.

Network Services and Software

TIP
If you want to test whether your computer's network card and network software are functioning properly, you can ping your IP address or ping 127.0.0.1, which is called the *loopback address*.

The second utility, called a **Trace Route**, also tests the connection to another computer on the Internet. The difference between Trace Route and Ping is that Trace Route will tell you every network device it has to go through to reach the destination. Try this utility to see how many "pit stops" are made before the trace gets to the final destination. Again, using the same Command window, type **tracert** followed by a space, and type the IP address of the computer you want to test.

Trace Route
A utility that is used to identify the path a packet has taken to reach a destination device on the Internet.

In the screenshot, you can see that the route taken to www.sierraclub.com started at alter.net in San Francisco and traveled through several routers until it reached the destination. It passed through a total of 13 routers on its trip.

Review Questions

Terms to Know
- authentication
- BSD UNIX
- central processing unit (CPU)
- domain
- fault tolerance
- hard disks
- Internet service provider (ISP)
- Linux
- multiprocessing

1. Why is multitasking important in a network operating system?

2. How does multiprocessing improve performance on a server?

3. What hardware is needed to create a fault-tolerant server?

4. Explain the function of a domain controller in a Windows 2000 network.

5. Why is NDS (Novell Directory Services) of interest to network administrators and others responsible for managing networks?

6. What are three advantages of using the UNIX operating system?

_____

_____

7. Why is there so much interest in the Linux operating system?

_____

_____

8. What are three reasons why IT administrators should consider using Mac OS X Server?

_____

_____

9. What command can be used to confirm IP address configuration for Windows XP, and what information does it provide?

_____

_____

10. Explain the difference between two tools that can be used to test an IP connection.

_____

_____

Terms to Know
- ❏ multitasking
- ❏ NetWare
- ❏ network operating system (NOS)
- ❏ Novel Directory Services (NDS)
- ❏ Ping
- ❏ pre-emptive multitasking
- ❏ processor
- ❏ Trace Route
- ❏ UNIX

Chapter 5

Topologies

A network's topology defines the way in which computers, printers, and other devices are connected. The topology you choose to implement will influence many factors in the way your network works.

When we refer to a network's topology, we may be referring to either its physical or its logical topology. A physical topology is a "map," or description, of the layout of the network media that interconnects the devices on a network. The physical topologies that are implemented today include the bus, ring, star, mesh, and a hybrid of several topologies combined into one network. A logical topology defines the way in which devices communicate and data is transmitted throughout the network. This may be in the form of a bus or ring.

In this chapter, you will learn:

   The difference between a physical and logical topology

   The characteristics of each physical topology

   How the two most common logical topology implementations, bus and ring, compare

   How the topology affects the network and its performance

   How wireless technologies allow users to communicate with wired network services

Physical versus Logical Topologies

media
The physical transmission pathway that connects devices in a network.

In networking terminology, the physical layout of a network's devices and **media** is the physical topology. The way in which data accesses media and transmits packets across it is the logical topology of a network. Here, the word *logical* can be misleading because it really has nothing to do with the way the network looks like it is functioning. Calling it the *logistical* topology would be closer to the truth—a logical topology is the way data travels the network, logistically.

At first glance, you may *think* you know the logical way in which the devices on your network are communicating, but be careful—it's not always as straightforward as it may appear. For example, you might think a network with cable running from a central wiring closet out to each device on the network would be labeled a physical star topology and a logical star topology, too. Unfortunately, it is not that clear-cut. Depending on whether the device that connects those cables in the closet is a hub or a MAU, the logical topology may be a bus or a ring—there is no logical star topology. Confusing? It won't be after reading this chapter.

The logical topology you choose will influence the physical layout of the cable—and vice versa. The decisions you make will be difficult to change later on. You need to consider both the physical and logical implications before you make any decisions about the type of network to install. For example, a physical star topology usually uses a logical bus topology. A physical ring topology typically uses a logical ring topology. The physical and logical topologies affect the way the cabling is installed, the network connection devices used, and the protocols used to transmit data.

TIP
The most important considerations in network design today are the potential to handle future expansion and bandwidth issues. If you choose the wrong topology in the beginning, you may have to reinstall the network media and purchase all new devices later.

Topologies

Types of Physical Topologies

Physical topologies are defined purely by the way in which the networking media connects the devices. A diagram of the physical topology of a network shows the path the media take to reach each of the devices on the network. The three most common physical topologies are bus, ring, and star:

physical topology
The physical layout of a network's media.

Ring

Bus

Star

The type of physical topology you choose for your network will affect how devices on your network communicate. Some of the factors to consider are:

- Cost
- Scalability
- Bandwidth capacity
- Ease of installation
- Ease of troubleshooting

Today, the majority of networks use a star, extended star, or hybrid topology. This allows for the implementation of either the Ethernet or Token Ring standard.

Bus Topology

A physical **bus topology** looks a lot like a bus line going through a city. The devices are connected by a single cable, which runs throughout the network. The main cable segment must end with a **terminator** that absorbs the signal when it reaches the end of the line. Without a terminator, the electrical signal that represents the data would reach the end of the copper wire and bounce back, causing errors on the network.

bus topology
A physical topology that utilizes a single main cable to which devices are attached.

terminator
A device used to terminate the ends of the main cable in networks implementing a physical bus topology and using Thinnet cabling.

coaxial cable
A networking media that consists of a single copper wire, surrounded by plastic and foil insulation, and covered with a plastic jacket.

BNC connector
Connector used with coaxial cable.

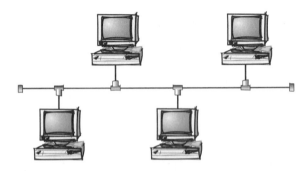

Networks that have a physical bus topology usually use thin **coaxial cable** that connects to the NIC using a **BNC connector**. As you can see in the previous illustration, this connector can be used to connect two or more computers together in a daisy-chain fashion, or a terminator can be added to one end of the cable.

BNC (bayonet connector) plug

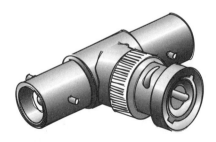

BNC T connector

Topologies

The Institute of Electrical and Electronics Engineers (IEEE) 802.3 standard, discussed in Chapter 10, has set the length of a cable segment and the number of devices on an **Ethernet** bus network. For a Thinnet network, also referred to as 10BASE2 and Thin Ethernet, the maximum segment length is 185 meters (about 600 feet), that connects a maximum of 30 devices. You may use up to four repeaters joining a maximum of five segments (three of which can be populated with devices), for a total cable length of 925 meters (3035 feet). The total number of **nodes** allowed is 150.

Another less-commonly-found option is Thicknet, which is also known as Thick Ethernet and 10BASE5. Thicknet cabling uses a thicker-gauge coaxial cable than Thinnet, and was a frequent choice for **backbone** installations. The cable segment limitation is 1,625 feet (500 meters), with a maximum of 100 nodes per segment. Thicknet allows a total of four repeaters for a total cable length of 8,125 feet, with a total of 300 nodes.

NOTE
It is important to follow the standards for network media so that the data packets can reach their destinations.

Ethernet
A network technology that incorporates Carrier Sense Multiple Access/Collision Detection (CSMA/CD) to access the media and uses a logical bus topology.

node
Any device connected to the network that is capable of sending and receiving data.

backbone
A single point of access, typically high-speed, for communication between multiple network segments.

ring topology
A physical topology in which all devices are connected in a circle, providing equal access to the network media.

Choosing a Bus Topology

The advantages of a bus topology are:

- ◆ Bus networks, Thinnet in particular, are inexpensive to install.
- ◆ You can easily add more workstations.
- ◆ Bus networks use less cable than other physical topologies.
- ◆ The bus topology works well for small networks (2–10 devices).

The disadvantages of a bus topology are:

- ◆ It's no longer a recommended option for new installations.
- ◆ If the backbone breaks, the network is down.
- ◆ Only a limited number of devices can be included.
- ◆ It's difficult to isolate where a problem may be.
- ◆ Sharing the same cable means slower access time.

Ring Topology

As its name implies, a **ring topology** is a topology in which the stations are connected in the form of a ring or circle (physical ring); in which the data flows in a circle, from station to station (logical ring); or a combination of both. It has no beginning or end that needs to be terminated. This allows every device to have an equal advantage accessing the media.

There are two kinds of ring topologies:

- Single ring
- Dual ring

When the first ring networks were installed, they used a single-ring topology, as shown in the following illustration. In a single-ring network, a single cable is shared by all the devices, and the data travels in one direction like a merry-go-round. Each device waits its turn and then transmits. When the data reaches its destination, another device can transmit.

Token Ring
A network access method that incorporates token passing to access the media. Data travels in only one direction in a Token Ring environment.

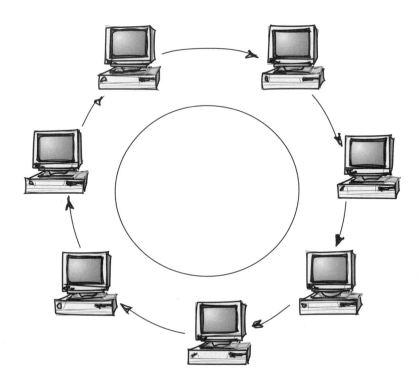

In a single-ring topology, the data travels in only one direction.

The most common implementation of the ring is in a **Token Ring** network. (Token Ring networks can be installed using a physical ring or a physical star. The 802.5 Token Ring access method is explained in Chapter 10.)

Topologies

As technology evolved, a dual-ring topology was developed. This topology allows two rings to send data, each in a different direction. Not only does this let more packets travel over the network; it allows packets to continue along the media, creating **redundancy**.

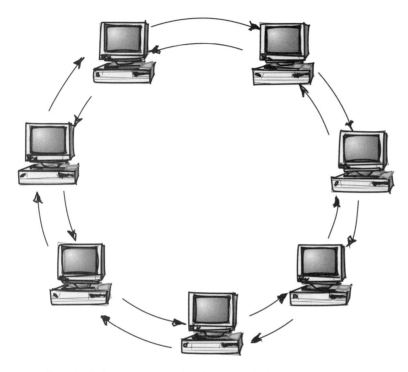

In a dual-ring topology, the data travels in two directions.

redundancy
The use of a secondary system identical to the primary that is capable of continuing service in the event that the primary system fails.

Fiber Distributed Data Interface (FDDI) is a technology similar to Token Ring, but it uses light instead of electricity to transmit data. FDDI networks use two rings for redundancy. As you will learn in Chapter 10, FDDI is unique compared to other types of ring networks because it will keep functioning in the event there is a break in one or both rings.

Choosing a Ring Topology

The advantages of a ring topology are:

- Data packets can travel at greater speeds.
- There are no collisions.
- It is easier to locate problems with devices and cable.
- No terminators are needed.

The disadvantages of a ring topology are:

- A ring network requires more cable than a bus network.
- A break in the cable will bring many types of ring networks down.
- When you add devices to the ring, all devices are suspended from using the network.
- It's not as common as the bus topology, so there's not as much equipment available.

star topology
A physical topology that connects networking devices to a central hub.

Star Topology

A physical **star topology** is installed in the shape of a star, like spokes in a bicycle wheel. As you can see in the following illustration, a star topology is made up of a central connection point, a **hub**, where the cable segments meet. Each device in a star network is connected to the central hub with its own cable. Although this does require more media, it has many advantages over both the bus and ring topologies.

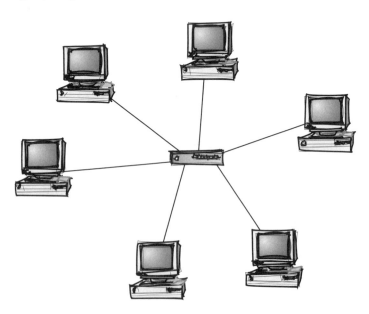

Star networks and extended star networks have quickly become the predominant topology type for most networks. One of the advantages of a star topology is that it's easy to make changes and additions to the network without disrupting users. You can add a new workstation or expand the network without ever affecting the network's performance.

Topologies

Choosing a Star Topology

The advantages of a star topology are:

- It's easy to add more devices as your network expands.
- The failure of one cable or one cable break will not bring down the entire network.
- The hub provides centralized management.
- It's easy to find device and cable problems.
- A star network can be upgraded to faster network transmission speeds.
- It's the most common topology, so there are many equipment options available

The disadvantages of a star topology are:

- A star network requires more media than a ring or bus network.
- The failure of the central hub can bring down the entire network.
- The costs of installation and equipment are higher than for most bus networks.

Extended Star Topology

If a star network is expanded to include an additional hub connected to the main hub, as shown in the following illustration, it is called an **extended star topology**.

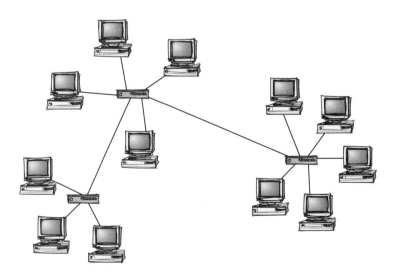

extended star topology
A physical topology that connects additional hubs to a central star topology to add additional devices.

Chapter 5

> **NOTE**
> The star network is the most commonly installed physical network topology today.

mesh topology
A physical topology in which all devices are interconnected. Also called a *net topology*.

fault tolerance
The capability to continue network operations despite problems or device failures.

Mesh Topology

A physical **mesh topology**, as shown here, is one that you should know about, though it is not very common in a local area network environment. It is most commonly used in wide area network connections to interconnect LANs. In a mesh topology, each device is connected to every other device. This allows all the devices to continue to communicate if one connection goes down. It is the ultimate in network interconnection security. In other words, it has **fault tolerance**.

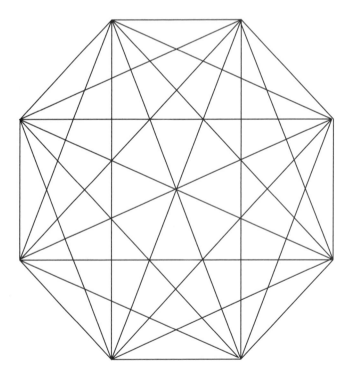

106

Topologies

Choosing a Mesh Topology

The advantage of a mesh topology is:

- A mesh network offers improved fault tolerance if part of the system goes down.

The disadvantages of a mesh topology are:

- It's expensive and difficult to install a mesh network.
- A mesh network is difficult to manage.
- A mesh network is also difficult to troubleshoot.

Wide area networks use a mesh topology between LANs to avoid system failures. The following illustration shows a mesh topology in a wide area network scenario:

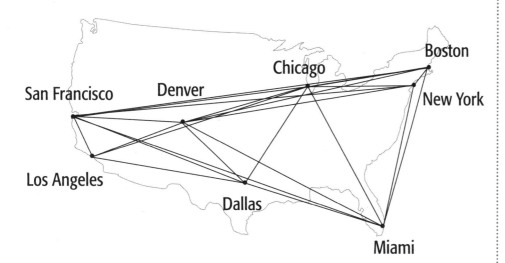

Chapter 5

Hybrid Topologies

Most networks implement a single topology, although that is changing. As old networks are updated and replaced, older segments may be left with a **legacy** topology. A **hybrid topology** combines two or more different physical topologies in a single network. The two most common hybrids found today are the star-bus and star-ring topologies.

When two hubs of different topologies are joined so that the devices attached to them can communicate, as in the following illustration, it is called a **star-bus** network.

legacy
From the past.

hybrid topology
A physical topology that combines one or more physical topologies, including the bus, ring, and star.

star-bus
A mixed physical topology that includes a star and a bus, and that is connected together using a device like a hub.

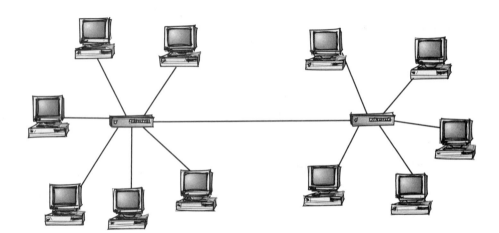

When two or more star topologies are linked together using a specialized hub called a MAU (you'll learn more about this device a little later), you have a star-ring topology:

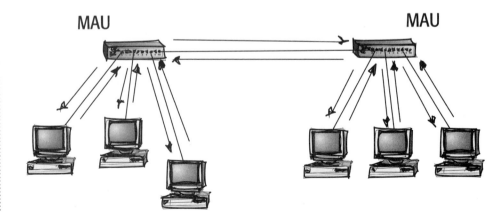

Topologies

Types of Logical Topologies

Now that you've learned all of the options for physically connecting the media and devices in your LAN, you need to understand the different ways in which the devices in a LAN communicate and transmit data. These are the **logical topologies**, and there are only two of them: bus and ring.

When a network is *physically* cabled in a bus, it is easy to understand that it is sending data in a *logical* bus throughout the network—that is, the data moves from device to device in a linear fashion. Likewise, a physical ring is easily interpreted as a logical ring. The following illustration compares the path the data takes in a logical bus network and in a logical ring network:

logical topology
The way in which data is transmitted throughout a network.

logical bus topology
A system in which data travels in a linear fashion away from the source to all destinations.

Logical bus vs. ring

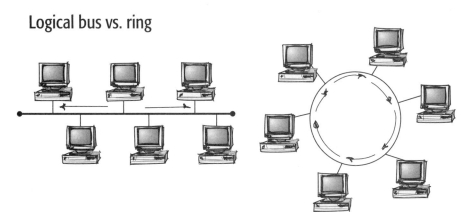

When a network uses a physical star topology, media can be accessed and data sent in either a logical bus or a logical ring; it depends on the network connection devices used, such as a NIC and a hub or MAU, and the layer 2 protocol that is implemented.

Logical Bus

In modern Ethernet networks, the physical layout is a star topology. At the center of the star is a hub. It is what happens inside that hub that defines the logical topology. An Ethernet hub uses a **logical bus topology** inside to transmit data to all the segments of its star.

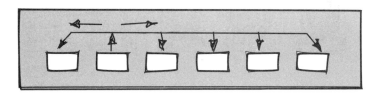

Chapter 5

The advantages of a logical bus topology are:

- ◊ If a node is down, it does not bring down the entire network.
- ◊ It's the most widely implemented of the logical topologies.
- ◊ Additions and changes can be made easily without affecting other workstations.

The disadvantages of a logical bus topology are:

- ◊ Collisions can occur easily.
- ◊ Only one device may access the media at a time.

Logical Ring

If the data on a network travels from one device to the next, in turn, then returns to the first device. We call this a **logical ring topology**. The most common implementation of the logical ring is the Token Ring network, in which data is transmitted using token passing. In token passing, a device receives an empty "token" that is being passed around the ring. If the token is empty, the device may fill that token with data and send it on around the network's ring. FDDI is another network technology that uses token passing.

It is also possible for a network with a physical star topology to transmit data in a logical ring. A device called a **MAU** (or MSAU, short for Multistation Access Unit) is the central device in this type of Token Ring network. The MAU, pronounced "mauw" (as in "mouse"), may look just like a regular hub. But as you can see following, inside the MAU the data is passed from device to device using a logical ring.

logical ring topology
A system in which data travels in a logistical ring from one device to another.

MAU (Multistation Access Unit)
The central hub in a Token Ring network that is wired as a physical star.

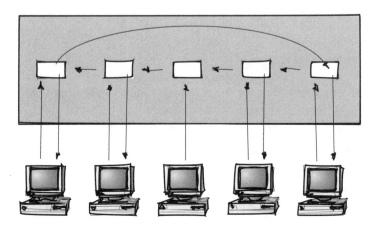

Topologies

The advantages of a logical ring topology are:

- The amount of data that can be carried in one message is much greater than on a logical bus.
- There are no collisions.

The disadvantages of a logical ring topology are:

- A broken ring will stop all transmissions.
- A device must wait for an empty token to be able to transmit.

Chapter 5

Wireless LAN Topologies

WLAN
Wireless Local Area Network.

AP
See *access point*.

access point (AP)
A device or software used to extend a wired LAN by transmitting wireless data signals to wireless nodes.

ad-hoc mode
A manner of transmitting wireless signals between devices or stations without the use of an access point. Also called *peer-to-peer mode*.

peer-to-peer mode
A manner of transmitting wireless signals between devices or stations without the use of an *access point*. See *ad-hoc mode*.

infrastructure mode
A manner of transmitting wireless signals that utilizes an access point as an extension of a wired network. This allows communication between wireless nodes or the wired network.

Wireless networks are not subject to the limitations of traditional networks and their topologies. Because the wireless transmission devices transmit in an omni-directional manner (in all directions), they are not limited to the traditional star, bus, or other topologies. In the following image, you can see three nodes communicating in **ad-hoc mode** using wireless network cards. This is also referred to as *peer-to-peer mode*.

Wireless LANs, **WLANs**, can also be an extension of a wired LAN. Devices communicate with each other and the wired network resources by using an **access point** (or **AP**). This is called the **infrastructure mode**. Access points can be combined in an overlapping fashion to ensure that signals transmit to any hard-to-reach places. Following is an example of the way two access points extend a LAN in infrastructure mode.

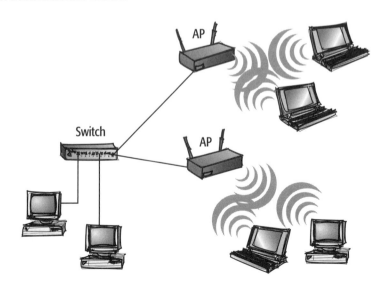

See Chapters 8, 9, and 10 for more information on wireless LANs.

Review Questions

1. Define the physical topology of a network.

2. Define the logical topology of a network.

3. What are the five most common physical topologies?

4. Which is the most common physical topology today?

5. What type of media is typically used with a physical bus topology?

6. Which network access method is usually implemented if a physical bus topology is used?

7. List one advantage and one disadvantage of installing a physical bus topology.

8. Describe the characteristics of a physical ring topology.

9. Which network access methods use a physical ring topology?

Terms to Know
- access point (AP)
- ad-hoc mode
- backbone
- BNC connector
- bus topology
- coaxial cable
- Ethernet
- extended star topology
- fault tolerance
- hub

Review Questions

Terms to Know
- hybrid topology
- infrastructure mode
- legacy
- logical bus topology
- logical ring topology
- logical topologies
- MAU
- media
- mesh topology

10. List one advantage and one disadvantage of installing a physical ring topology.

11. Describe the characteristics of a physical star topology.

12. Which network access methods can be implemented using a physical star?

13. Describe a physical mesh topology.

14. List one advantage and one disadvantage of installing a physical mesh topology.

15. What are the two logical topologies used in networks?

16. Which logical topology or topologies can be implemented with a physical star?

17. What physical and logical topologies can be implemented with Ethernet?

18. What physical and logical topologies can be implemented with Token Ring?

19. What are the physical limitations of wireless LANs?

_____

20. Define ad-hoc or peer-to-peer mode in wireless LANs.

_____

21. Define infrastructure mode in wireless LANs.

_____

Terms to Know
- nodes
- physical topology
- redundancy
- ring topology
- star topology
- star-bus topology
- star-ring topology
- terminator
- Token Ring

Chapter 6

Electricity

Without electricity, we would not have LANs, MANs, or WANs—or even a light by which to plug in our machines. In fact, electricity plays a very important role in network design and implementation: For one, the pulses that create the data signals that are sent over copper media are electrical. Networks also require sufficient power to run, and this requires planning. Finally, a surge in electrical signals, natural or man-made, can cause interference within networks that can lead to loss of data.

When planning and designing a network, you must consider all the issues related to electricity. It is the responsibility of the designer, as well as the network administrator, to make sure protection devices are budgeted for and put in place. Monitoring power fluctuations can help an administrator accommodate network needs and plan appropriately for providing protection. To keep the network functioning properly, it's crucial that equipment is put in place that keeps things running in the case of either excess power or power loss.

To help you understand electricity and its impact on the network, this chapter will explore:

   Two types of electricity

   The problems associated with excess power and their solutions

   The problems associated with a reduction or loss of power, and their solutions

   Proper grounding

Chapter 6

Types of Electrical Currents

There are two types of electrical currents that can be used by electrical devices. In this section, you'll learn about:

- ◊ Alternating current
- ◊ Direct current

alternating current (AC)
An electrical current that reverses its direction at regularly recurring intervals.

electron
A negatively charged particle.

sine wave
A waveform that represents the positive and negative changes in an AC electrical current.

generator
A machine that converts mechanical energy into electrical energy.

transformer
A device that uses induction to convert a primary circuit into a different voltage in a secondary circuit.

Alternating Current

Alternating current (AC) is the type of current we get from the power company. The term comes from the fact that **electrons** in the wire alternate the direction in which they move. As the current flows from positive to negative, the voltage rises and falls. The following illustration shows what an AC current looks like when it is measured as a **sine wave**.

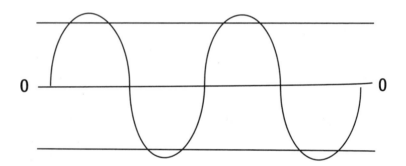

At the power plant, electricity is created by **generators** and sent to our homes and businesses. In order for the electricity to reach our homes, the power company must force the electricity from the plant out onto the wires.

Electricity is forced onto the line using **transformers**. Transformers raise the electricity's force, which is called *voltage*. High-voltage transformers are found at power plants and at the substations from where the high-voltage electricity is carried via transmission lines to your neighborhood. In each neighborhood, there are small transformers mounted high up on utility poles where the electricity is

Electricity

converted from a high voltage to a level that is safe for use in your home. Unlike the lines you have in your home, high-voltage lines carry thousands to millions of volts and are raised up by transmission towers.

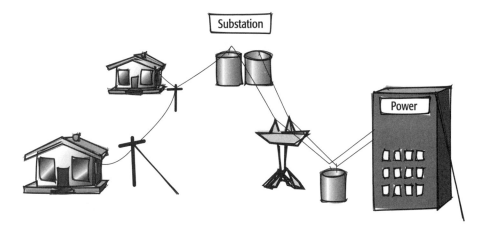

As you can see, high-voltage lines run to a substation in which a transformer lowers the voltage so that it can be sent to your home, school, or office building. From the substation, power runs to your home through power lines called *distribution lines*. You may see distribution lines in your neighborhood running along poles. They also run underground.

At the power plant and substations, as well as where the electricity enters your home, there are circuit breakers. These circuit breakers stop the flow of electricity if there is a dramatic increase in current caused by lightning or some other surge. In addition, circuit breakers stop the flow of electricity when there is excessive demand on the **circuit**. You may have "popped" a circuit breaker in your house by using too many kitchen appliances at the same time.

In addition to electrical wires, there are also wires that are connected to metal (typically copper) rods, planted in the ground under or near a building. When the wires are connected to these rods, it gives excess electricity a place to go should there be too much voltage for the line. The ground then absorbs the electricity. These wires are called *earth grounds*.

circuit
The path for the usage of electricity.

NOTE

Lightning rods used in areas where lightning is common are separate, extended earth grounds used to attract and displace excess current.

Earth Grounds

Earth grounds are very important at both the power plant and your home. Every home built today has an earth ground. The ground wire is connected to the wires in your home and to the ground plug in your outlet.

earth ground
A wire that allows excess electricity to go into the ground. It has a potential electrical charge, which measures zero. Also called a *ground*.

hot wire
The electrical wire in an outlet that is the source of the incoming electrical signal from the power company.

neutral wire
The electrical wire in an outlet that provides the return path for the electrical signal, which completes the circuit. It has a ground potential of zero.

normal mode
Electrical power problems between the hot and neutral wires in an electrical circuit or outlet.

common mode
Electrical power problems between the hot or neutral wires and the ground wires in an electrical outlet.

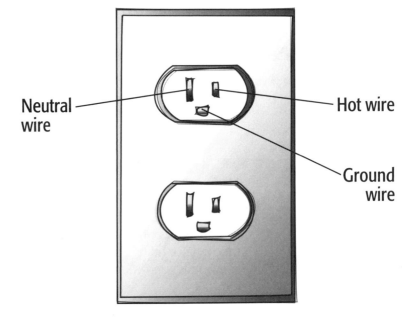

When you plug a device into an outlet and turn it on, a circuit is formed between the hot wire and the neutral wire. The **hot wire** is the wire through which the electricity from the plant travels; the **neutral wire** is the wire through which the electricity continues to flow.

Problems between the hot and neutral wires are called **normal mode** problems; they are usually caused by surges from an internal source, such as load switching within the building; or from an outside source, such as lightning. Problems between the hot or neutral wire and the ground wire are called **common mode** problems; they are usually caused by electrical noise from wiring faults or overloaded power circuits.

Not all electrical outlets are the same. Electrical outlets installed before the 1960s do not usually have a ground wire. Outlets in other countries may be very different from ours, and may even carry different voltages. Even some newer homes have been built without a properly connected ground. To find out if you have a true grounded power outlet, you must use a tester like the one shown here. This tester will show whether the outlet is correctly wired and grounded; and if there is a problem, it will show which wire is at fault.

Electricity

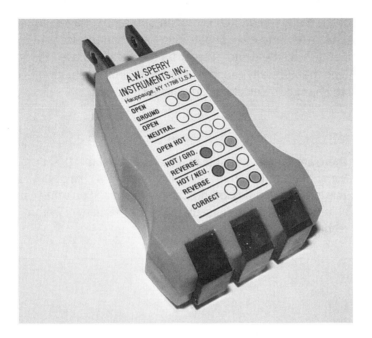

direct current (DC)
An electrical current that flows in one direction only and is substantially constant in value.

TIP
This device is a must for every network administrator's toolbox. It should be used before installing any electrical device to check the circuit.

Direct Current

Direct current (DC) is the type of current we find in batteries, which can power a device without it having to be "plugged-in." When the circuit is completed, direct current provides a direct and constant current. With DC current, power does not fluctuate from positive to negative, as happens with AC current. The following illustration shows a DC sine wave:

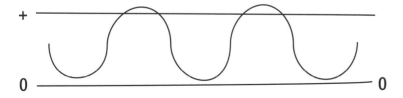

Like AC power, battery power can come in different voltages—for example, you can buy A, AA, AAA, and even D batteries. Each battery size provides a different amount of energy to the device. Larger batteries can be used for devices, such as computers, that typically run from AC power. In this situation, a converter is necessary to change the electrical flow from DC to AC.

NOTE

There are various rechargeable batteries available that store direct current. Many of them lose their capability to store a charge over time, however. This effect, called *shadowing*, can be minimized on some types of rechargeable batteries by completely using up the battery before beginning the recharging process.

Electricity

Power Problems and Solutions

Computers and electrical devices are sensitive to changes in the electrical current they receive. There are several types of electrical problems that can damage electrical equipment on your network and disrupt service. These problems can be broken down into three categories:

- Static electricity
- Excess power
    - Spikes
    - Surges
- Reduction or loss of power
    - Sag/brownout
    - Blackout

static electricity
The electrical energy produced by friction.

electrostatic discharge
The displacement of static electric buildup.

Static Electricity

Static electricity is a common by-product of our presence in a world full of electrically charged electrons. By walking across a carpet or even brushing against nylon cloth, you can charge your clothing and yourself to the point of damaging computer equipment and data. Although you cannot prevent static electricity from forming, you can prevent it from damaging your networking equipment.

When you work on electrical equipment, you need to be sure you are always correctly grounded to avoid passing on any static that has accumulated in your body. You can do this by wearing a wrist strap. When the wrist strap is attached to your wrist, a metal contact is made with your body. An insulated wire runs from the strap to an alligator clip that is attached to the metal case of the computer. When you make contact with the computer, the wrist strap allows an **electrostatic discharge**.

You should always displace any electrostatic charge in your body by touching a grounded metal device such as a metal computer cabinet or case. Remember, it is only grounded while it is plugged in.

Excess Power

Electrical equipment, especially the equipment found on computer networks, is very susceptible to electrical damage. Excess power on the network presents a serious danger to a network. When the electrical power that supports equipment increases dramatically, it can translate into lost data. The data loss can be temporary, as is the case when a computer must be turned off and restarted. Data loss can also be more severe, causing permanent, unrecoverable damage to equipment.

Chapter 6

The two most common problems associated with excess power signals are:

- Surges
- Spikes

These electrical problems are typically not visible to the human eye until a major problem occurs. Surges and spikes can be monitored with special software. But monitoring these power problems only provides evidence as to their existence. You still need to understand how to prevent them.

A more-subtle problem that is caused by excessive electrical power is the electrical interference that can affect network media. As you know, when data travels on network media such as Category 5 UTP cabling, it does so in the form of electrical signals. These signals are very sensitive to interference. Excess power from the electrical wiring can interfere with the data being sent. When network media is placed too close to powerful electrical circuits, the electromagnetic field of the electrical wires can change the pattern of the data signals on the media. If the change is significant, the network devices will not be able understand the data. Usually, these types of problems, called electromagnetic interference (EMI), can be extremely difficult to find and correct.

Surges

A **surge** is a short-term increase, typically at least 1/120th of a second, in an electrical signal of at least 110 percent of the normal voltage. A surge can be the result of increased electrical flow from the power company or a natural cause, such as lightning. This electrical problem is the most common of the excess-power problems, and is the most damaging to network equipment.

Consider a typical computer or network device that is running at 120 volts, which is the typical voltage found in the U.S. During a surge, the power directed at the device is now more than 132 volts. This is a big shock that can easily damage sensitive electronic equipment.

If you were to graph a surge in an electrical signal, it would look like this:

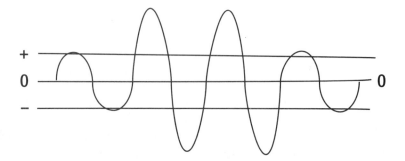

---

surge
An increase in the electrical signal to 110 percent or more of the normal voltage for less than 2.5 seconds.

Electricity

WARNING
Hubs are very vulnerable to electrical damage and are often overlooked when planning for surge protection.

Spikes

A **spike** is a significant increase in electricity on a line that lasts for a very short amount of time. Typically, a spike will last from less than 1 **microsecond** up to 100 microseconds. The graph here shows a spike of 240 volts, or twice the typical power voltage.

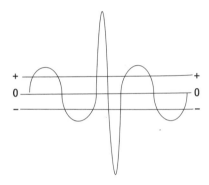

Protecting Data and Equipment

Although you cannot prevent electrical surges and spikes from happening, you can prevent them from damaging your electronic devices. A **surge suppressor** that contains a **metal oxide varistor (MOV)** will protect your devices from many if not all the problems caused by excess power.

Surge suppressors provide a barrier between the device and the power source, absorbing or redirecting excess energy. These devices can be built into the circuit-breaker box or wall outlet, or used between the wall outlet and the device's power plug. If a spike or surge reaches the suppressor, it will be redirected to the ground wire or absorbed before it damages the device.

TIP
Surge suppressors are rated to withstand specific amounts of excess energy. Most can redirect up to 330 volts. Be sure to buy one that has a high rating (stated in **joules**).

---

spike
An increase in the electrical signal of at least twice the normal voltage for 10 to 100 microseconds.

microsecond
One-millionth of a second.

surge suppressor
A device that is used to absorb excess electrical signals to prevent damage to electrical devices.

metal oxide varistor (MOV)
An electrical component of a surge suppressor or line conditioner that absorbs excess electrical signals.

joule
A unit of energy equal to the work done by a force of one newton acting through a distance of one meter.

Chapter 6

All devices on the network should use surge suppressors to prevent damage from electrical power surges. Examples of surge suppressors are shown in the following graphic:

uninterruptible power supply (UPS)
A device that converts stored direct current to alternating current to provide power in the event of a reduction or loss in the primary electrical signal.

sag
A reduction in electrical signal to 80 percent of the normal voltage. Also known as a *brownout.*

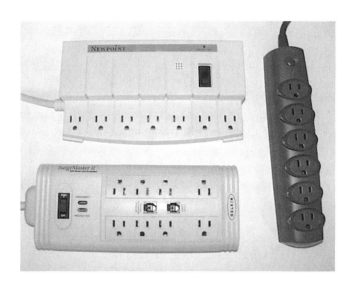

Power Loss

When a computer or other related networking device experiences a loss of power, data may be lost and the device may even be harmed. There are two main types of power loss that you should be aware of:

- Sags or brownouts
- Blackouts

Sags and blackouts are more noticeable to the human eye than surges or spikes. Flickering lights or monitors indicate that a sag is occurring.

You can prevent power-loss problems by installing an **uninterruptible power supply (UPS)** that will provide backup power to a device. These devices are discussed in Chapter 13.

Sags

If the network experiences a decrease in electrical power to 80 percent or less of the normal voltage, it is called a **sag**. In the United States, where the normal voltage is 120 volts, this would mean a drop to 95 volts.

Electricity

A sag can be caused by a large device, such as an elevator or even a microwave oven, drawing a large amount of power from the line and taxing the total power being provided at that moment. You have seen this happen in a building when lights flicker or go dim when someone turns on a toaster oven or the air conditioner kicks on.

Although sags are not as damaging as many other power-related problems, they can cause a computer to shut down or reboot, possibly causing a loss of data or files you are working on. They can even cause the loss of directories on the server. Prolonged exposure to sags can be especially damaging to equipment.

Blackouts

A complete **blackout** shuts down all electrical devices and leaves you in the dark. Blackouts can damage devices as well as the work on your computer. They are very common in all environments.

Just about everyone has experienced the frustration of having a blackout shut down their computer before they've saved their most recent work. Blackouts also can cause extreme damage to servers, which require proper shutdown. When a server is powered off abruptly by a blackout, you will also lose any data on the wire from unaffected devices that are saving their files to the server's disk.

TIP

Power conditioners, or **line conditioners**, are devices that are installed between a building's power source and the devices being protected. They increase or decrease voltage to regulate the power that is being delivered.

---

blackout
A complete loss of electrical signal.

power conditioner
A device that helps to regulate electrical power and provide protection to electronic devices. Also known as a line conditioner.

line conditioner
A device that helps to regulate electrical power and provide protection to electronic devices. Also known as a *line conditioner*.

Chapter 6

Data Centers

Large technology installations that serve businesses and organizations with data processing and business services are often called **data centers**. They are unique environments that support the special needs of sensitive electronic equipment. Some of the requirements associated with data centers are:

Data center
A facility that supports large numbers of computers and other technology.

- $24 \times 7 \times 365$ availability
- Network redundancy
- Power redundancy
- Security
- Environmental control
- Monitoring

Data centers require special considerations and equipment, including fail-safe power. It is important to plan ahead for all the power needs that will be required. Considerations in planning a data center include:

- Power requirements
- Adequate cooling
- Sufficient space
- Raised flooring

TIP
100 watts per square foot is a common starting place for planning electrical needs within a data center. Many provide for double that amount.

Review Questions

1. What are the two types of electrical currents we use to provide power to our devices?

   _____

   _____

2. What is alternating current?

   _____

   _____

3. What is direct current?

   _____

   _____

4. What type of electricity is created by friction caused by our movements?

   _____

   _____

5. Why is static electricity harmful to devices?

   _____

   _____

6. What can you do to prevent an electrostatic discharge from damaging your devices?

   _____

   _____

Terms to Know
- ❑ alternating current (AC)
- ❑ blackout
- ❑ circuit
- ❑ common mode
- ❑ data center
- ❑ direct current (DC)
- ❑ earth ground
- ❑ electrons
- ❑ electrostatic discharge

Review Questions

Terms to Know
- generator
- hot wire
- joule
- line conditioner
- metal oxide varistor (MOV)
- microsecond
- neutral wire
- normal mode

7. What types of problems can result from excess electrical current in the line?

8. What is a surge?

9. What is a spike?

10. What can you do to prevent excess electrical current from damaging your devices?

11. What types of problems can result from a reduction in electrical signal?

12. What is a sag?

13. What is a blackout?

_____

_____

14. What can you do to prevent a loss of data or damage to devices as a result of sags or blackouts?

_____

_____

15. What is a line or power conditioner used for?

_____

_____

16. What are some of the special considerations that should be addressed when planning a data center?

_____

_____

Terms to Know
- ❏ power conditioner
- ❏ sag
- ❏ sine wave
- ❏ spike
- ❏ static electricity
- ❏ surge
- ❏ surge suppressor
- ❏ transformer
- ❏ uninterruptible power supply (UPS)

Chapter 7

Signaling

In computer communications, information is transferred using any number of media, such as electricity, light, or radio waves. In the case of most networks, electrical signals are used to transfer a document to a printer or connect to a web page on the Internet. The signals used for data communications can come in many different forms, though their purpose—transmitting information—is always the same.

In this discussion of signaling, you will learn about the following topics:

   What signals are

   How electrical and radio signals are measured

   The differences between analog and digital signals

   Digital-to-analog and analog-to-digital conversion

   The types of transmissions used in networking

   How transmissions flow over media

Chapter 7

What Is a Signal?

On traditional computer networks that use cables, an electrical signal is represented as a change in voltage over time. The change occurs when the voltage begins at 0 and increases until it peaks. The signal then drops back down to its lowest level. Other types of signals also fluctuate from high to low levels.

interference
Also referred to as *electromagnetic interference (EMI)* or *radio frequency interference (RFI)*, interference is a disturbance in the electromagnetic field of a circuit that may make the signal distorted or unrecognizable.

If signals are distorted, communication may be interrupted or stopped all together. You sometimes experience signal distortion, or **interference**, when on a telephone call or when using a mobile phone. During your conversation, you hear static or aren't able to hear anything through the receiver. The interference may last for only a second, and then you are able to continue with your conversation. If the interference doesn't last very long, you can probably fill in what was lost.

Computers and computer networks are not as forgiving when it comes to signal interference. The signal needs to be clear and in an understandable format, or else the computer won't be able to convert the signal into data. The computer treats bad signals as electrical garbage, throwing out whatever it doesn't understand.

On more modern networks, it is common to have network transmissions occurring in part using wireless devices. Unlike wired networks that rely on electrical signals, wireless networks use radio waves. The wireless devices transmit and receive data by locking in on the same channel. If they both know the channel to tune it to, they can pick up the signal.

In order to design, install, and manage computer networks, you need to understand signaling. As you have learned, cables and radio waves represent the roads that information travels on, but they do not generate signals. Network cards, hubs, and other devices are responsible for sending and receiving electrical signals. The signals are then transformed into data as the information is passed up the layers of the OSI model.

NOTE
See Chapter 2 for a review of the layers of the OSI model and the process of encapsulation.

When troubleshooting network problems, you may have to use a cable tester to test for signal quality on wired networks and software on a laptop for wireless networks. If the signal quality is poor on a cabled network, there may be an elevator or another electrical device such as a computer, monitor, or heater

Signaling

near the network cable causing interference. Wireless networks can also be adversely affected by large electrical machinery, other devices that produce strong radio signals, and solid obstructions like concrete walls.

NOTE

You will learn about cable testers in Chapter 13.

Chapter 7

Measuring Signals

When a network card transmits data across the network, it sends out a signal that fluctuates in **voltage**. The pattern that the signal makes is called a **waveform**. The two types of waveforms are sine waves and square waves:

voltage
The electromotive force that moves electrical current against resistance.

waveform
The shape of a signal that has been graphed according to its amplitude and frequency during a specified period of time.

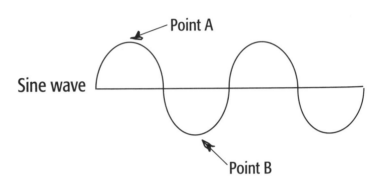

amplitude
The maximum value of a signal as measured from its average state, which is usually zero.

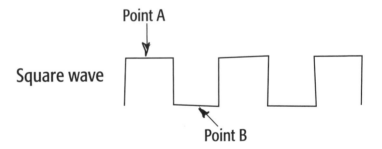

ampere
The unit of measurement of electrical current. Also called an *amp*.

frequency
The number of cycles completed in one second.

Both waveforms shown previously fluctuate from a high signal strength (point A) to a low or zero signal strength (point B). The maximum value of the signal strength, point A, represents the signal's **amplitude**. Another characteristic of a waveform is its **frequency**. Frequency is measured by adding the number of wave patterns completed in a given period of time. In the following diagram, the wave pattern has repeated three times in one second:

hertz (Hz)
A unit of measurement for the frequency of cycles in a signal.

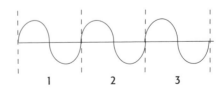

The number of times the waves cycle is measured in **hertz (Hz)**. The electricity coming out of a socket in your home cycles at 60Hz (60 cycles per second).

Signaling

Signals and Computers

By now, you have probably purchased a computer or helped someone purchase a computer. As you know, one of the biggest decisions you make when purchasing a computer is deciding whether it is worth paying the additional money for a faster one. Most people measure the performance of a computer on the speed of the central processing unit (CPU), also known as the *processor*. Although this is not the only component affecting computer performance, it is an important one. Some processors are able to process instructions at speeds as high as 3GHz, which is equivalent to 3 billion hertz, or cycles, per second. As you will learn later, that can translate into the potential for a computer and a network to transfer a lot of information.

Analog Signals

Earlier, we showed you two waveforms, a sine wave and a square wave. **Sine waves** are used to represent the volume (amplitude) and pitch (frequency) of an analog signal. The most common type of analog signal is sound or the human voice. When we speak, we are using air to transmit an analog signal. Sound traveling over the air is not the only form of **analog signal**. Electrical signals, like music from an audiotape, can also be in analog form.

sine wave
A curve representing fluctuations from high to low that recur at a constant amplitude.

analog signal
The fluctuation of an electrical signal or other source that has the characteristic of being continuous rather than having discrete points, as is the case with digital signals.

AM and FM Radio

An example of analog signals are the radio signals used by AM and FM radio stations. Have you ever wondered why you can't hear the radio in the air? Well, it's because humans are unable to hear radio signals because their frequency is too high for the human ear to distinguish.

The human ear can only recognize sounds with frequencies of 20KHz or less. **AM (amplitude modulation)** radio operates between 535KHz and 1605KHz. **FM (frequency modulation)** radio operates between 88MHz and 108MHz. (Notice that these numbers represent the range of stations on your radio.) Radio stations want to be on the FM band because FM signals can transmit more clearly. With more cycles per second, more information can be transmitted.

amplitude modulation (AM)
Signals in the same frequency that have different amplitudes.

frequency modulation (FM)
Signals that may have the same amplitude, but have different frequencies.

digital signal
The transmission of data using a discontinuous source, which can be accomplished with electricity or electromagnetic fields.

square wave
The waveform used to represent digital bits in their electrical form.

binary
A mathematical system that uses base 2 rather than the decimal numbering system that uses base 10.

decimal
A numbering system that uses base 10.

Amplitude modulation

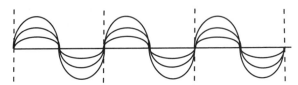

Three AM radio stations all running at the same frequency but different amplitudes

Frequency modulation

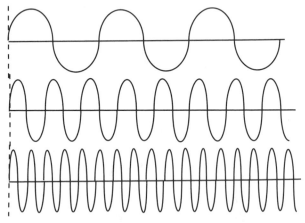

Three FM radio stations all running at different frequencies

Digital Signals

When computers store data or communicate, they are using **digital signals**. Digital signals are graphically represented using the **square wave**. The reason the wave is "square" has to do with the way computers use information. In computers, all information, including numbers, letters, and colors, is represented in **binary** form. Binary data consists of 1s and 0s. A combination of 1s and 0s makes up a single character. The binary number 0001 is equal to 1 in **decimal** form.

Signaling

Each 0 or 1 is called a **bit**. A bit is the most basic form in which data can be presented in a computer. A bit of information moves through a computer or computer network as an electrical signal. As you can imagine, transmitting an entire text document involves thousands of bits. When the electrical signals that represent the bits are being transmitted, it is called a **bit stream**.

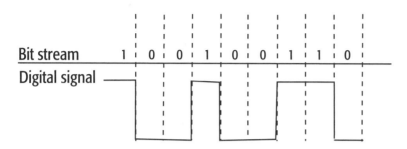

bit
The smallest form of information in a computer that has a value of either 1 or 0.

bit stream
Describes the transmission of a contiguous group of bits in a digital signal.

bandwidth
The total capacity of a connection or transmission measured in bits.

Garbled Digital Signals

Unlike analog signals, digital signals must be transmitted clearly and accurately for the receiving computer to be able to interpret the electrical signal. If the signal is garbled due to interference of some kind, the receiving computer ignores the information. The computer doesn't understand anything but 1s and 0s; therefore, a signal that cannot be deciphered by the computer as a 1 is interpreted as a 0. The change of a bit from a 1 to a 0 equates to garbled data.

Intended Bits	1	0	1	1	0	0
Actual Bit Stream	1	0	0	0	0	0
Digital signal						

Digital Signals and Bandwidth

On a network, you want to be able to transfer as much data as possible. The theoretical maximum amount of data transferred per second is called **bandwidth**. Bandwidth is measured in bits per second (bps).

Following is a list of the different measurements of bandwidth and the amount of data that can be transferred per second for each:

1 bit per second (bps)	= 1 bit per second
1 kilobit per second (Kbps)	= 1,000 bits per second
1 megabit per second (Mbps)	= 1,000,000 bits per second
1 gigabit per second (Gbps)	= 1,000,000,000 bits per second

Bits versus Bytes

Bits are not the same as bytes. A byte is equal to eight bits. Bytes are used to represent the size of a file, storage space, or memory. The following table shows how bytes convert into bits:

```
1 byte = 8 bits
1KB    = 1,024 bytes     = 8,192 bits
1MB    = 1,048,576 bytes = 8,388,608 bits
```

If the frequency of a digital signal is 8Hz, then the bandwidth is 8 bits transferred per second, or 8bps (slow by all standards). Most networks today are capable of transferring data at 100Mbps, because the frequency of the digital signal operates at 100MHz.

WARNING

The actual bandwidth of a network is often slower than what is advertised. A 10Mbps network sounds great on paper, but in the real world the network might only perform at 37 percent of its capacity, or 3.7Mbps. New network devices like switches, discussed in Chapter 9, improve on this problem by dedicating 10Mbps or 100Mbps to each connection. If you are considering building or upgrading a network, a switch should be the obvious choice.

On faster networks that operate at 100MHz and 1GHz, the margin for error shrinks. You already know that computers are picky when it comes to sending and receiving digital signals. As the speed of the network increases, more and more cycles are sent in one second, which makes it more likely that bits will get lost.

Signaling

Think of a conveyor belt in a bottling factory. As the bottles come off the conveyor belt, someone has to put the caps on. If the conveyor belt goes too fast, some bottles may not get caps. Just like a bottle can't be sold without a cap, a computer can't accept a partially transmitted signal. The challenge on high-speed networks is to fit in more cycles, millions more, in the same amount of time as the slower networks.

noise
Another term for *interference*.

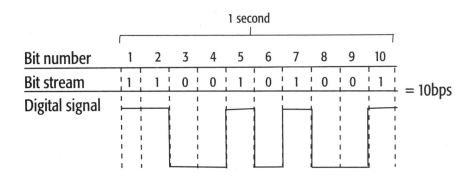

Analog versus Digital

It is important to understand the basic differences between analog and digital signals. It is obvious that digital signals are important because they are used for computer networking. Analog signals, although not used on computer networks, can be as important—for example, to organizations that need cable television in some locations.

The advantages of analog signals are:

- They're best suited for transmissions of audio and video.
- Analog signals consume less bandwidth than digital signals to carry the same information.
- Analog systems are readily in place around the world.
- Analog is less susceptible to **noise**.

The advantages of digital signals are:

- They're best suited for transmission of computer data.
- Digital data can easily be compressed.
- Digital information can be encrypted.
- Equipment that uses digital signals is more common and less expensive.
- Digital signals can provide better clarity because all signals must be either 1s or 0s.

Chapter 7

An Analog *and* Digital World

The vast analog infrastructure that is in place to transmit voice communications has become the ideal system to use for data communications. Many businesses and homes around the world have access to telephone lines. The telephone system provides a reliable way for people to connect to the Internet over either slower connections, usually less than 56Kbps using modems, and faster connections using more modern digital signals over the same lines.

In order for computers to communicate over analog phone lines, a conversion process needs to take place. A modem, the name of which is derived from the terms *modulator* and *demodulator*, converts a digital signal from the computer into an analog signal that is sent over a telephone line.

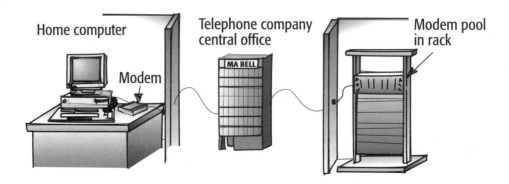

Unlike modem technology, which uses the existing voice infrastructure of telephone companies, new digital technologies have found their way in to every part of our lives. Using the same telephone lines that are used for voice signals, telecommunication companies discovered decades ago that they could send multiple analog signals over a single telephone line. It wasn't implemented when it was first discovered because of the costs. Now, that innovation means that a digital signal for data transmission can be sent over the same copper wire as a telephone line. The same is true for cable television.

Signaling

NOTE

In addition to these wired technologies, wireless networks have become one of the fastest growing areas in networking. In addition to wireless local area networks (WLANs), personal wireless networks and mobile devices have taken what were analog-only signals like radios and walkie-talkies, and spawned a completely digital wireless age. Unlike AM and FM radio signals, discussed earlier in this chapter, wireless networks operate at the 2.4GHz and 5GHz range. Because of these high frequencies, wireless networks can transmit even larger amounts of data then before. There are several wireless standards that have been ratified since 1997. The original wireless standard had a bandwidth range of 1Mbps. The latest wireless standard in development has an expected range of 75Mbps. Wireless standards are reviewed in Chapter 10.

Chapter 7

Understanding Transmission

Computer signals define the form of the electrical current as it is sent across a network. The way a signal is transmitted is a different matter. The transmission mode for a digital signal on a computer network describes the way that two devices communicate.

Transmission Types

Now that you understand how to connect to the Internet using a modem, it is important to know how transmission occurs. When data is transferred between computers, there are two different transmission types: asynchronous and synchronous. Both types are responsible for letting the receiving computer know when data begins and ends during a transmission, a process known as **bit synchronization**.

Asynchronous Transmissions

Asynchronous communication is a transmission type used by modems. In asynchronous communication, bit synchronization between two modems is made possible using start and stop bits. The start bit indicates the beginning of the data, and the stop bit indicates the end. Each character that is transmitted has a start bit and stop bit. Because extra bits are added to the data being transmitted, more bandwidth is consumed.

The characteristics of asynchronous transmissions are:

- ◊ They're less efficient than synchronous transmissions due to overhead of extra bits.
- ◊ They're best suited to Internet traffic in which information is transmitted in short bursts.
- ◊ They're less resistant to disruption.

Synchronous Transmissions

Synchronous transmissions don't make use of start and stop bits to maintain bit synchronization. Instead, synchronous communication is established when two devices agree on the duration of a transmission of a bit, called **timing**.

---

bit synchronization
A function that is required to determine when the beginning and end of the transmission of data occurs.

asynchronous
A transmission type that uses start and stop bits for bit synchronization.

synchronous
A transmission type that uses timing for bit synchronization.

timing
A function used in synchronous transmissions, in which two systems that are communicating each agree on the time that it takes to transmit one bit of data.

Signaling

The characteristics of synchronous transmissions are:

- They transmit faster than asynchronous transmissions because there is no excess overhead due to the transmission of start and stop bits.
- Data is transmitted in blocks rather than one character at a time.
- Synchronous transmissions slow down on poor-quality lines.

Transmission Modes

Now that you understand how data is transmitted, you can look at ways that the transmission signals flow over media. The most common medium on which electrical signals are transmitted is twisted-pair cable. Twisted-pair cable consists of pairs of thin copper wires twisted together, encased in a common casing or sheath. Each end of the cable is connected to network equipment.

The telephone system also uses twisted-pair cable. At one end, the wires connect to the telephone company equipment. At the other end is the telephone in your home. The connection of the two wires at each end creates a circuit. And it is the circuit that allows telephone calls to be transmitted between your home and the rest of the telephone system.

Simplex Transmissions

In **simplex** transmissions, information is sent in only one direction. There is no mechanism for information to be transmitted back to where the original signal came from. A loudspeaker system at a sports stadium is an example of simplex transmission. An announcer speaks into a microphone, and their voice is sent through an amplifier and then out to all the speakers. Many fire alarm systems work the same way.

simplex
A transmission flow that allows communication in only one direction.

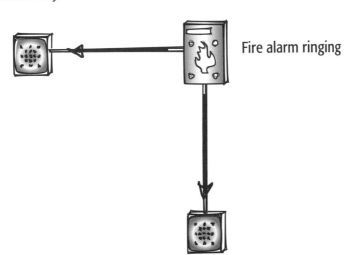
Fire alarm ringing

Half-Duplex Transmissions

In **half-duplex (HDX)** transmissions data is sent and received between two devices, but not simultaneously. A walkie-talkie, for example, can only send or receive a transmission at any given time. It can't do both.

half duplex (HDX)
On a computer network, data travels in either direction, but not at the same time.

In the illustration of a half-duplex transmission that follows, computer A sends information to computer B. At the end of the transmission, computer B sends information to computer A. Computer A cannot send any information to computer B while computer B is transmitting data.

full duplex (FDX)
On a computer network, data can travel in both directions simultaneously.

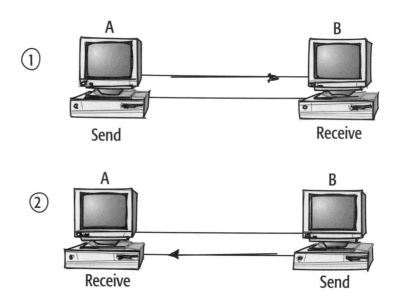

Full-Duplex Transmissions

Full-duplex (FDX) transmissions have one big advantage over half-duplex transmissions, in that transmissions can occur simultaneously in both directions. On a typical Ethernet network using twisted-pair cable, the bandwidth of the network is 10Mbps. If the network hardware is capable of full duplex, the network can be fully utilized. Data can be sent at 10Mbps and at the same time can be received at 10Mbps. Sometimes full duplex is mistakenly described as a way of doubling the network speed. The truth is that the maximum throughput of data being sent is still 10Mbps. The network just transmits faster because there is no pause between sending and receiving. It is important to remember that to have full-duplex transmission, both network devices that are connected to each other must support it. That includes network interface cards, hubs, and other network equipment.

Signaling

Send and receive simultaneously

TIP

Network devices that support full duplex are usually set to auto-negotiate the duplex type when they are shipped from the manufacturer. Half or full duplex is configured using software.

WARNING

Remember that devices at both ends of the connection have to support full duplex. Even then, if the network devices aren't from the same company, full duplex may not work, or may cause errors on your network.

Review Questions

Terms to Know
- AM (amplitude modulation)
- amplitude
- analog signal
- asynchronous
- bandwidth
- binary
- bit
- bit stream
- bit synchronization
- decimal
- digital signals
- FM (frequency modulation)
- frequency

1. The waveform that a digital signal makes is called a _____.
    A. Sine wave
    B. Digital wave
    C. Square wave
    D. Inversion wave

2. The number of cycles per second is measured in _____.

3. Match the bandwidth in bits with the correct unit of measurement:
    A. 1Kbps
    B. 1bps
    C. 1Gbps
    D. 1Mbp

    _____ 1,000,000 bits per second
    _____ 1,000 bits per second
    _____ 1,000,000,000 bits per second
    _____ 1 bit per second

4. When a series of bits is transmitted as a digital signal, it is called a(n) _____.

5. Why is it important to have a clear digital signal?

6. Explain the purpose of a modem and why modems are important.

7. Place an A in front of the items that describe characteristics of asynchronous transmissions and an S in front of the items that describe characteristics of synchronous transmissions.

    _____ Data is transmitted in blocks of bits.

    _____ Performance is not as good on poor-quality lines.

    _____ It's less efficient due to overhead of extra bits.

    _____ It transmits faster because of no excess overhead.

    _____ It's best suited to traffic that occurs in short bursts.

8. Which type of transmission flow allows data to be transmitted in both directions, but only one way at a time?

    A. Simplex
    B. Half duplex
    C. Full duplex

Terms to Know
- full duplex (FDX)
- half duplex (HDX)
- hertz (Hz)
- interference
- noise
- simplex
- sine waves
- square wave
- synchronous
- timing
- voltage
- waveform

149

Chapter 8

Network Media

A network's transmission media is a long-term investment. Unlike computer equipment, which is often replaced every two to five years, a company may use the same networking media for 10 or 15 years. Thus, choosing the correct media is crucial to having a functioning network.

There are many types of media that can be used to carry transmissions across a network to allow end devices to communicate. Common types include copper, glass, and air. Each type of transmission media has characteristics that will affect the network's performance.

To provide a solid understanding of the different types of media, this chapter will cover:

   The types of networking media

   The characteristics of each type

   The connectors used with each type of media

   The advantages and disadvantages of each media type

   The distance limitations of each media type

Chapter 8

Network Media and Connectors

There are many network **media** types, including traditional copper cable, glass, and air. The vast majority of existing networks use some form of copper cabling. The types of copper cable in use include coaxial cable, shielded twisted pair, and unshielded twisted pair. For higher-speed networks, many organizations are electing to use fiber-optic media, although copper can provide exceptional transmission rates—even into the Gbps range. There are also networks that use no cable at all. These wireless networks transmit data over air. Wireless networking allows you to transmit to places where installing a physical cable would be impossible. It also allows for mobile networking and provides an opportunity to easily implement changes in the physical layout of end devices.

> **media**
> The material used as a means of transmission.

Almost as important as the network cable choice is the type of connector that attaches to the ends of the cable to interconnect devices. Connectors are used in all types of media, including wireless networks. The type of connector used depends on several factors, the most important of which are the type of media used and the type of network. This chapter will explain the following types of networking connectors: BNC, RJ-45, SC and ST, SMA, FDDI, and the IBM types.

Media is possibly the single most important long-term investment you will make in your network and should be chosen with great consideration. Your choice of media type will affect the type of NICs installed, the speed of the network, and the capability of the network to meet the needs of the future. If the network media can't support data transfer rates beyond the speed of the network's current NICs—say, 10Mbps—then in the future that network will have difficulty transmitting video, voice, or other bandwidth-intensive data.

Although currently fiber-optic cable is considered to be the ideal network media because of its high bandwidth potential, it can be prohibitive in terms of cost. However, Category 5e UTP cable is able to transmit at speeds over 1Gbps.

Today, a Category 5 cable with RJ connectors and a 10/100Mbps NIC may be purchased for $25.00. A fiber-optic NIC, which was rarely heard of not long ago because the cost was so prohibitive, can now be found for less than $200.00. A few years ago, few people believed there would be a need to run fiber-optic cable to the desktop, but now in some offices it is considered the de facto media type. Companies are seeing greater potential in paying now for fiber-optic cable to take advantage of faster network technologies in the future.

Network Media

Copper Media

Copper cable is the oldest form of networking media. It was implemented in the first IBM mainframe networks using shielded twisted pair. Today, there are several types of copper media that provide different data transmission rates and serve different types of networks. These include coaxial cable, shielded twisted pair, and unshielded twisted pair.

Coaxial Cable

Coaxial cable—*coax* for short—was the first type of networking media used in Ethernet LANs. From ARCNET to Ethernet, many different network types utilized some kind of coaxial cable. There are still many networks that are using coaxial cable.

Coaxial cable is constructed from one central copper wire that is covered with a plastic insulator, called a **dielectric**, and shielded from interference by a foil wrapping or braid. The outer jacket that protects the cable is made from plastic and can be either PVC or **plenum**. The covering on plenum cable is made from Teflon and is a fire-retardant. The foil or braided shielding in the coaxial cable must be grounded to protect against electromagnetic interference.

coaxial cable
A type of media that consists of a single copper wire surrounded by insulation and a metal shield of foil or braid, and covered with a plastic jacket made of PVC or plenum.

dielectric
A type of insulation material.

plenum
A type of plastic that meets fire safety standards. It does not burn or create toxic fumes when exposed to heat or fire.

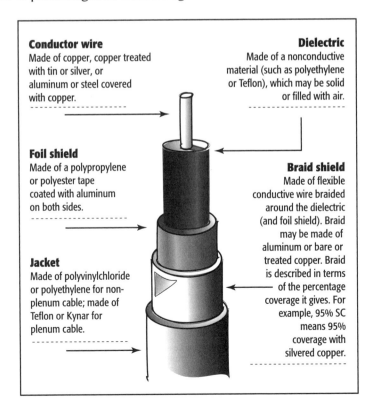

Conductor wire
Made of copper, copper treated with tin or silver, or aluminum or steel covered with copper.

Dielectric
Made of a nonconductive material (such as polyethylene or Teflon), which may be solid or filled with air.

Foil shield
Made of a polypropylene or polyester tape coated with aluminum on both sides.

Braid shield
Made of flexible conductive wire braided around the dielectric (and foil shield). Braid may be made of aluminum or bare or treated copper. Braid is described in terms of the percentage coverage it gives. For example, 95% SC means 95% coverage with silvered copper.

Jacket
Made of polyvinylchloride or polyethylene for non-plenum cable; made of Teflon or Kynar for plenum cable.

153

Types of Coaxial Cable

There are many types of coaxial cable that are used in networking and other cable types used for other types of transmissions, such as cable television. Each type of coax has specific characteristics that meet the needs of the type of transmission being carried. The following table shows the most common types of coaxial cable, comparing their **impedance** in **ohms**, the types of networks that implement them, and some of their basic characteristics. Many types of coax use the RG (radio grade) classification, which defines the size of the copper center and the diameter of the outer jacket.

impedance
The opposition to electrical current, measured in ohms.

ohm
A measurement that translates to one unit of resistance.

Classification	Impedance	Implementation	Description
RG-6	93 ohms	Cable TV	Like RG-59, but larger in diameter to accommodate higher bandwidth for cable TV transmissions
RG-8	50 ohms	Thick Ethernet (Thicknet)	Solid copper core approximately 0.4 inches in diameter; must use a drop cable to attach to a device's NIC
RG-11	75 ohms	Cable TV	Often has four layers of shielding
RG-58/U	50 ohms	Thin Ethernet (Thinnet)	Solid copper core less than 0.2 inches in diameter
RG-58 A/U	50 ohms	Thinnet	Like RG-58/U, but with a stranded copper core
RG-58 C/U	50 ohms	Thinnet	Same as RG A/U, but used in military applications
RG-59	75 ohms	ARCNET; cable TV	Thick copper core with a large outer housing
RG-62	93 ohms	ARCNET; IBM mainframes	Used in IBM 3270 systems to connect terminals to the mainframe

Impedance, measured in ohms, defines the resistance to electrical current. Higher impedance means higher resistance.

Network Media

Thicknet and Thinnet

In the past, two types of coaxial cable were commonly used in networking:

- Thicknet
- Thinnet

Thicknet is no longer installed, but it may exist in older networks, so you should be familiar with its characteristics. Thicknet was the original cabling used in Ethernet networks. It ran through walls or ceilings, and often used a connection device known as a **vampire tap** to connect devices to the main cable through the use of a drop cable.

The following illustration shows a common Thicknet cable with a vampire tap and a **drop cable**. Thicknet is often called 10Base5. The 10 refers to the 10Mbps bandwidth of the cable, whereas *Base* indicates that the transmission type is a **baseband** transmission. Thicknet is extremely difficult to install and work with because of the unusual size and stiffness of the cable.

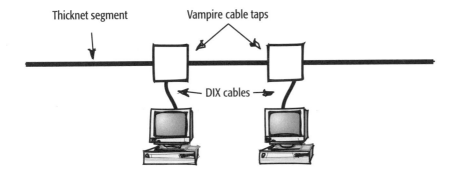

TIP
Although you might inherit a Thicknet network, you shouldn't consider installing one. There are much better alternatives that cost less, are easier to install, and perform better—such as Category 5 UTP cable.

Thinnet, also known as 10Base2, was developed after the implementation of Thicknet. Thinnet is smaller in diameter than Thicknet and is much easier to work with. The connectors used with Thinnet, known as BNC connectors, can be either barrel connectors that attach one cable to another or T connectors that attach devices to one cable. These connectors simply have to be connected and twisted into place to ensure their connection. They can be used to implement a bus network that is easy to add devices to but difficult to troubleshoot.

Thicknet
A type of coaxial cable, typically used in older networks, that is approximately 0.4 inches in diameter. Also called *10Base5*.

vampire tap
The type of connector used to attach a drop cable to a Thicknet backbone cable.

drop cable
A portion of cable that attaches from an end device to a vampire tap attached to a Thicknet cable.

baseband
Transmissions that occur using a single fixed frequency.

Thinnet
A type of coaxial cable, typically used in older networks, that is approximately 0.2 inches in diameter. Also called *10Base2*.

Chapter 8

Should You Choose Thinnet?

The advantages of using Thinnet coaxial cable are:

- It's easy to install.
- It's small in diameter.
- Its shielding, when grounded, reduces **EMI** and **RFI** interference.

The disadvantages of using Thinnet are:

- If a cable breaks, the entire network goes down.
- The cable must be grounded to prevent interference.
- It's more expensive than unshielded twisted pair.
- The connectors can be expensive.
- It does not support high-speed transmissions.

Coaxial Cable Implementations

Today, Thinnet coaxial cable is sometimes, although rarely, installed in networks. It provides a simple way for a small number of users to create a network using a physical bus topology. It is widely available in computer and office supply stores.

NOTE

See Chapter 5 to learn the advantages and disadvantages of a bus network.

Thinnet is considered antiquated by many, but its ease of implementation and low cost are appealing. Many NICs have both a coaxial cable connector and an unshielded twisted pair (UTP) connector, allowing you to change media as the network develops and the coaxial cable is phased out.

TIP

Terminating coaxial cable with a BNC connector is much more difficult than using Category 5 UTP cable and RJ-45 plugs. It must be properly connected to ensure grounding of the metal shielding.

EMI (electromagnetic interference)
When the electric field created by electricity on one wire changes the form of an electrical charge on an adjacent wire.

RFI (radio frequency interference)
Interference with transmissions over copper wires caused by radio signals.

Network Media

Shielded Twisted-Pair Cable

Shielded twisted pair (STP) is made up of pairs of copper wires that are twisted together. The pairs are covered in a foil or braided shielding, as well as an outer PVC jacket. As with coaxial cable, the shielding must be grounded to prevent the foil or braided shielding from becoming a magnet for electricity.

shielded twisted pair (STP)
A type of media that consists of pairs of twisted wires that are insulated by foil and covered in a plastic jacket.

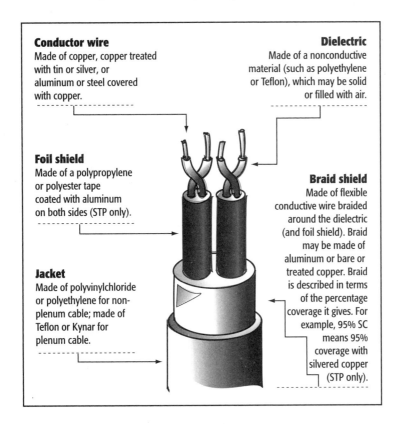

Conductor wire
Made of copper, copper treated with tin or silver, or aluminum or steel covered with copper.

Dielectric
Made of a nonconductive material (such as polyethylene or Teflon), which may be solid or filled with air.

Foil shield
Made of a polypropylene or polyester tape coated with aluminum on both sides (STP only).

Braid shield
Made of flexible conductive wire braided around the dielectric (and foil shield). Braid may be made of aluminum or bare or treated copper. Braid is described in terms of the percentage coverage it gives. For example, 95% SC means 95% coverage with silvered copper (STP only).

Jacket
Made of polyvinylchloride or polyethylene for non-plenum cable; made of Teflon or Kynar for plenum cable.

TIP

The color codes for 150 ohm STP-A cable are red and green wires for pair 1, and orange and black wires for pair 2.

Chapter 8

Types of Shielded Twisted Pair

There are five types of STP used in the LAN environment, as defined by IBM. (STP is most commonly referred to by the IBM categories, although it is used in ARCNET networks. It is rarely used in Ethernet networks.) The following table compares the types of STP cabling, the number of pairs within each cable, the **gauge** of copper used, and their implementation.

> **gauge**
> The measurement of the diameter of electrical wire.

Although considered less susceptible than UTP to interference, STP is subject to near-end **cross talk** and electromagnetic interference. Cross talk happens when neighboring wires interfere with other signal transmissions.

> **cross talk**
> The electromagnetic interference that occurs when the electrical signal on one wire changes the electrical properties of a signal on an adjacent wire.

Type	Number of Pairs	Gauge of Copper Used	Implementation
Type 1	2	22	Used in IBM's Token Ring networks for the main ring or to connect nodes to the MAU
Type 2	4	22	Used as a hybrid cable for voice and data
Type 6	2	26	Used as an adapter cable to connect a node to a MAU in Token Ring environments; used for shorter distances; can be used as a patch cable
Type 8	2	26	Uses flat wire and is designed to run under carpets; prone to signal loss but adequate for short distances
Type 9	2	26	Used between floors of a building; has a solid or stranded core and a plenum jacket

A wire's gauge indicates its thickness. The higher the gauge, the smaller the wire.

TIP
Although the IBM data connector is used in the IBM environment, STP cabling can also use the same RJ connectors that UTP uses.

Network Media

The connectors developed for use with IBM cabling are known as the **IBM data connectors**. They are square, genderless connectors that are designed to interconnect with one another. This photo is of a single data connector.

IBM data connector
The type of connector used in IBM networks with shielded twisted-pair cabling.

Should You Choose Shielded Twisted Pair?

The advantages of using shielded twisted-pair cable are:

- Shielding reduces interference and cross talk.
- It can be used with RJ connectors, which are common and inexpensive, instead of the IBM genderless connectors.

The disadvantages of using shielded twisted-pair cable are:

- It must be properly grounded.
- It's more expensive than unshielded twisted pair.
- It's difficult to terminate.

STP Implementations

Although STP used to be more frequently used, it is no longer commonly implemented in the IBM Token Ring environment and has been replaced by Category 5 UTP. It is rarely utilized in Ethernet networks, due to its cost and difficulty with termination. Terminating STP for IBM Token Ring networks takes time, due primarily to the fact that the cable is bulky and awkward, and the cable has to be grounded.

Chapter 8

Unshielded Twisted Pair

Unshielded twisted pair (UTP) is the most common implementation of copper media today. It is made from twisted pairs of color-coded copper wires, but does not include foil or braiding as insulation to protect against interference. Instead, the wire pairs within each cable have varied amounts of twists per foot to produce **cancellation**. Cancellation is important in UTP cabling because it is the only means of preventing interference. The following illustration shows how twisted pairs use cancellation to prevent interference from their neighboring wires:

unshielded twisted pair (UTP)
A type of media that consists of one or two pairs of twisted wires and is covered in a plastic jacket.

cancellation
The process in which two wires are twisted together to prevent outside interference or cross talk.

Underwriters Laboratories (UL)
UL is responsible for testing the safety of electrical wires including the types of cable used in networks.

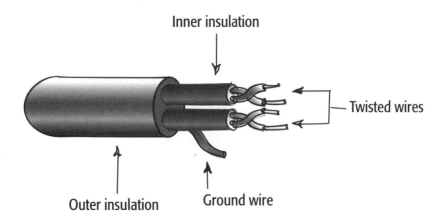

There are several types of UTP cables, identified by their category number. This identification should be marked on the plastic jacket, as required by the **Underwriters Laboratories (UL)**. The following photo shows the category marking on UTP cable:

Network Media

TIP
UTP cable is often referred to by *cat* rather than *category*. Cat 5 refers to Category 5 UTP cable.

In July of 2002, the **EIA/TIA** approved the Category 6 standard called 568-B2.1. This standard was designed to support applications running at 1 **gigabit** per second or better. Prior to that time, only four types of UTP—Category 3, Category 4, Category 5e, and Category 5—were specified in the EIA/TIA 568-A standard. It is likely that Category 6 will become the new standard for LAN installations. Category 5e cable is the most common type of UTP cabling currently being used for data networks. Backbone UTP has more than two pairs of wires, usually in multiples of 25.

Categories of Unshielded Twisted Pair

The table below shows the UTP categories, the number of pairs in each, the grade of cable each uses, and how they are implemented:

Type	Number of Pairs	Transmission Rate	Implementation
Category 1	2	Voice grade	Used in the telephone industry, but not suitable for data transmissions (though it has been used for short distances)
Category 2	2	4Mbps	Can be used in data communications, but is rarely installed; no longer recognized under the 568-A standard
Category 3	4	10Mbps	Used for 10Base-T networks and for voice communication
Category 4	4	16Mbps	Used in IBM Token Ring networks
Category 5	4	100Mbps and higher	Used in Ethernet and 100BaseX networks
Category 5e	4	100Mbps and higher	Used in Ethernet and 100–1000 Base-X networks
Category 6	4	100Mbps and higher	Used in Ethernet and 1000Base-X networks

gigabit
One billion bits.

EIA/TIA
Electronics Industries Alliance/Telecommunications Industry Association.

Chapter 8

NOTE

As of July 2002, Cisco was still recommending Category 5e as the installation standard, but not Category 6.

RJ-45
The type of connector used with twisted-pair cabling.

TIP

EIA/TIA standards are described in Chapter 10.

Installing and Terminating UTP Cable

When using UTP cabling, you need to consider the type of cable and connectors to use. Solid UTP, which consists of solid wires of copper twisted together, is used for permanent installations, such as within walls or raceways. These permanent cable runs, referred to as *horizontal cross-connects*, are terminated at one end at a patch panel in the wiring closet and at the other end with, typically, an **RJ-45** jack and wall plate. The following photo shows a variety of RJ-45 jacks by different vendors. Jacks can be purchased for either the 568-A or 568-B standard, though some provide the color order for both standards within one plug.

UTP stranded cable consists of thin, hair-like strands of copper within each individual wire. This type of UTP is more flexible than solid UTP. It is used for patch cords that connect devices to the RJ-45 jacks in the wall. Within the wiring closet, a patch cord is used between the patch panel and a hub or switch to connect the devices. These patch cords are terminated using RJ-45 plugs. These plugs look a lot like phone plugs (which are R-11 plugs), although they are slightly larger.

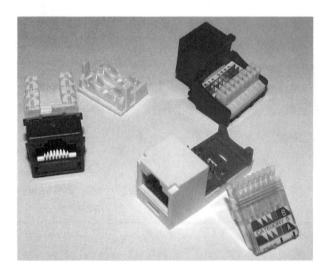

Network Media

Should You Choose Category 5 UTP?

The advantages of using Category 5 UTP are:

- It's inexpensive to install.
- It's easy to terminate.
- It's widely used and tested.
- It supports many network types.

The disadvantages of using Category 5 UTP are:

- It's susceptible to interference.
- It's prone to damage during installation.
- Distance limitations are often misunderstood or not followed.

NOTE

It is expected that Category 6 installations will be more cost-effective than Category 5.

UTP Implementations

UTP cabling is currently being installed in 80% of today's LANs. Although fiber-optic cabling is most commonly used as the backbone cable between network segments, UTP is installed as the media to support connections to desktop computers and other devices. Currently, special Category 5 UTP cabling, Enhanced Cat 5 or Cat 5e, can be certified to 350Mbps, which is 35 times as fast as what was being used only five years ago. The standard certification for Category 5 UTP is 100Mbps.

WARNING

When planning the installation of a network, make sure that the media will provide for the needs of the network for 10 to 15 years. It should have the longest life span of any part of your network. Don't skimp on it!

Fiber-Optic Media

Fiber-optic media offers several advantages over copper media and only a few disadvantages. Although many existing networks only utilize a fiber-optic cable as the backbone media, more companies are installing fiber-optic cable to the desktop for their LANs.

laser
An extremely focused beam of light. The term comes from "light amplification by simulated emission of radiation."

> **NOTE**
>
> Glass is the typical material used as the core of fiber-optic cable, although some companies are using high-grade plastic for the core. Plastic fiber is not currently accepted by the EIA/TIA 568-A standard.

light-emitting diode (LED)
Electrical signals converted into light by a semiconductor.

The use of fiber-optic media in a network prevents EMI and RFI from damaging the data signals. Fiber optics are resistant to the effects of lightning and electrical surges that can travel through copper media, resulting in the damage and destruction of devices. Fiber optics can also support a higher bandwidth for longer distances without the use of a repeater. Unlike copper media, fiber-optic media transmits data using light. The light source may be a **laser** light or a **light-emitting diode (LED)**. Although laser is the preferred source for light in a fiber-optic network, LED is more common. The cost for LED devices is lower, and their expected life is longer. LEDs are also more reliable.

fiber-optic cable
A type of media that uses very thin glass or plastic to transmit light signals.

Fiber-Optic Cable

micron
A measurement equal to 1/1,000,000 of a meter or 1/25,000 of an inch.

Fiber-optic cable is constructed of a center core of silica, extruded glass, or plastic. It is designed to pass specific types of light waves over long distances with very little signal loss. The center core of a multimode fiber-optic cable has a diameter of 125 **microns** (125μ), and is approximately the size of two human hairs.

cladding
A reflective material that surrounds the fiber-optic strand for the purpose of focusing the light down the fiber-optic strand and preventing the loss of light.

The following illustration depicts a cross-section of a multimode fiber-optic cable. Outside the center core is **cladding**, which is reflective material that helps bend the light waves as they travel down the cable. Kevlar, the material used in bulletproof vests, is typically used as a buffer and strength member to protect the glass fibers. Kevlar is extremely strong and resistant to damage, which helps strengthen the glass fiber.

Network Media

> **NOTE**
> Fiber-optic cables can be purchased as single fibers or as pairs of up to 36 fiber strands. Pairs of fiber are needed to complete a full-duplex circuit.

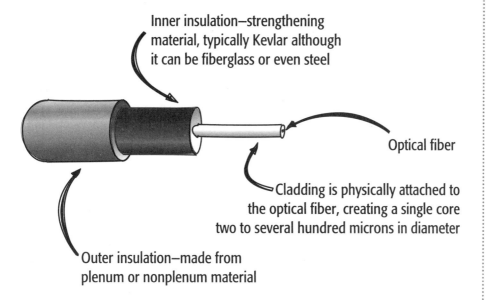

multimode
Fiber-optic cable that supports multiple transmission signals using LED or laser light.

single-mode
Fiber-optic cable that has only one cable and allows a single transmission signal at one time.

Fiber-optic cables can come in various sizes that are identified by the size of the center core. The most common size for fiber-optic cable is 62.5/125 microns. This is the typical core size for **multimode** fiber-optic cable. **Single-mode** fiber is more commonly used in long-distance applications, and has a core of 8/125 microns. Single-mode fiber can be implemented over three kilometers without needing a repeater. It can also be used in many LAN architectures and applications, including Ethernet, 10Base-F, FDDI, Optical Token Ring (Token Ring using fiber-optic cable), and some ATM architectures.

The following table shows the characteristics of single-mode and multimode fiber-optic cabling:

Cable Type	Core Size	Implementation/Characteristics
Single-mode fiber	8/125 microns	Used where large bandwidth is needed; can support only a single transmission at a time
Multimode fiber	62.5/125 microns	Most common fiber-optic cable; can be used in most network applications; allows multiple transmissions to occur at the same time

Fiber-Optic Connectors

Fiber-optic cabling may be installed with a number of different terminations. The choices include:

- SMA (SMA 905/SMA 906) connectors
- ST (straight tip) connectors
- SC (subscriber connector) connectors

SMA Connectors

SMA connector
Fiber-optic connector that has a screw-on connector.

The **SMA connector**, illustrated following, is a screw-on connector of which there are two different types. The 905 type has a straight connector. The 906 type has a connector that is smaller at the end so that two connectors can be joined with the use of a coupler. The SMA connector has lost the popularity it once enjoyed. The difficulty in pairing two connectors to meet the needs of the dual-fiber interface has limited the use of the SMA connector in many new installations.

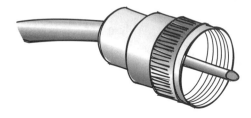

ST Connectors

ST connector
Fiber-optic connector that uses a twist-and-lock mechanism.

The second type of fiber-optic connector is the **ST connector**. It uses a connection similar to that of the BNC connector. After inserting the connector, it is twisted and locked into place. The ST is quick and simple to install, and for that reason it's popular. It is also difficult to pair in dual-fiber installations. The following photo shows a couple of ST connectors.

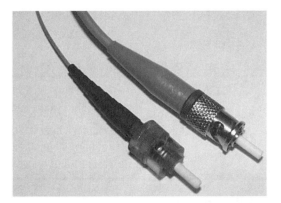

Network Media

SC Connectors

The third—and these days most popular—connector is the **SC connector**. The SC connector, unlike the SMA or ST connector, is easy to recognize with its square tip. The SC connector is easy to install and is often coupled into duplex cables that form sets. A paired set may include keying, which prevents it from being installed incorrectly. In a keyed set, one connector is shaped slightly differently, as is its mate on the other end. In this way, the fibers are not crossed at installation.

The keyed set SC connector that is shown following has been adopted under EIA/TIA 568-A as the recommended fiber connector. It is called the 568-SC.

SC connector
Fiber-optic connector that is square in form and can be keyed.

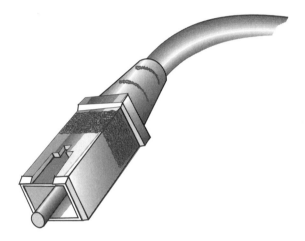

Should You Choose Fiber-Optic Cable?

The advantages of using fiber-optic cable are:

- It can be installed over long distances.
- It provides large amounts of bandwidth.
- It's not susceptible to EMI or RFI.
- It cannot be tapped easily, so security is better.

The disadvantages of using fiber-optic cable are:

- It's the most expensive media to purchase and install.
- Strict installation guidelines must be met for the cabling to be certified.

NOTE

Fiber-optic cable that will be used outdoors must be specifically designed for that application and have the appropriate conduit.

Chapter 8

TIP
The SC connector is the connector of choice in today's fiber-optic Ethernet networks.

TIP
Terminating fiber-optic cabling used to require extensive training, but today new techniques allow novices to learn how to terminate fiber in a matter of minutes. These new kits and fiber tips use specialized cutters and epoxy to ensure success.

Network Media

Wireless Networking Options

You know those giant antennas you sometimes see perched on top of tall buildings? Although they may be sending and receiving radio or television transmissions, they could also be transmitting data. Wireless communications are being installed all over the world to solve the problem of continuous communication, no matter where you are. WLANs, wireless local area networks, are increasingly being installed in offices and schools as a substitute for the traditional wired connections to each node.

By using high-frequency signals, wireless transmissions can cover large distances. They can share the air in the same way that radio and television signals are spread throughout the world. Each signal uses a different frequency—measured in hertz or cycles per second—so that they remain unique from one another. The following illustration shows how signals at different frequencies create completely different signal shapes and can be recognized by devices "tuned-in" to those frequencies. Transmission frequencies are explained in detail in Chapter 7.

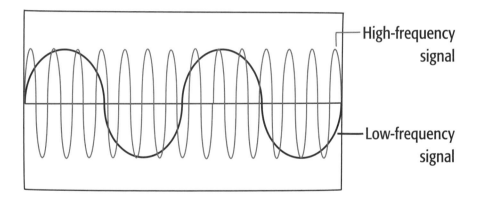

TIP

The standards for wireless data transmissions are described in IEEE 802.11, which was adopted in 1997. There are currently several standards for 802.11, including 802.11a, 802.11b, and 802.11g.

Today, wireless communications exist in LANs, MANs, and WANs. Because signals are transmitted through the atmosphere, this is a valuable medium in situations where cabling is not cost-effective. Depending on the implementation, the choice of wireless technology varies greatly. There are three basic types of wireless communications:

- Radio waves transmit at 10KHz to 1GHz.
- Microwaves transmit at 1GHz to 500GHz.
- Infrared transmits at 500GHz to 1THz.

Radio Transmissions

Radio waves are used to transmit and receive data in a way similar to that used by your local radio station: You tune in to the frequency that is being transmitted by the station you want to receive data from, and conversely, the other site must do the same to receive the data that you are sending. Radio transmissions used in networking use higher frequencies. Wireless LANs utilize radio transmissions to send and receive data.

There are a variety of types of radio transmission used in networks, including:

- Narrow-band
- High-powered
- Frequency-hopping spread spectrum
- Direct-sequence-modulation spread spectrum

Narrow-Band Radio Transmissions

One type of single-frequency radio transmissions used in wireless LANs, narrow-band transmissions, are unregulated by the Federal Communications Commission (FCC) and operate at these frequencies: 902–928MHz, 2.4GHz, and 5.72–5.85GHz. Narrow-band radio communications are limited to 70 meters, or 230 feet. This type of communication is extremely susceptible to interference, and can be picked up by anyone tuned into the right frequency, which may present a security concern when transferring data.

High-Powered Radio Transmissions

The other single-frequency radio transmissions used in wireless LANs, high-powered radio transmissions, have a line-of-sight limitation. A line-of-sight limitation is when the transmitting antennas must be in direct view of each other. These transmissions are more reliable than narrow-band transmissions but more costly to implement. Often, this type of transmission requires a license from the FCC. Typically, when this type of transmission is required, the service is purchased

Network Media

from a large communications company. The users who benefit most from high-powered radio transmissions are mobile users. When mobile users are out of range of the radio signal, a repeater tower can be employed to extend the transmission distance.

Spread-Spectrum Radio Transmissions

Unlike the single-frequency transmissions previously mentioned, spread-spectrum radio transmissions utilize multiple frequencies at the same time. This makes it more difficult to tap into someone else's communications.

Spread-spectrum communications fall into three categories, depending on how they utilize the multiple frequencies. The first type uses **frequency hopping**, **FH** or **FHSS**. FHSS switches the transmission of data between frequencies based on a timing pattern. Devices communicating on this type of network must be synchronized, and must also communicate between each other the next frequency that is going to be used. This type of transmission is extremely slow, transmitting at 2Mbps or less.

The second type of spread-spectrum transmission is **direct-sequence**, **DS** or **DSSS**. It uses several frequencies at the same time to transmit data. The data is broken into segments called **chips**. The capability to transmit at several frequencies at the same time makes for faster transmissions. Direct-sequence modulation networks can transmit at speeds from 2Mbps to 6Mbps. The devices on this type of network know how to reassemble the various chips back into the data segment that is being transmitted.

The third type of spread-spectrum is **orthogonal frequency division multiplexing**, **OFDM**. OFDM divides the channel into subchannels that are transmitted simultaneously at different frequencies to the receiver.

> **NOTE**
> 802.11a transmits in the 5**GHz** range using **orthogonal frequency division multiplexing**, or **OFDM**. It can send data at up to 54Mbps! This is a much higher transmission rate than the 11Mbps for 802.11b.

> **NOTE**
> OFDM was selected as the basis for the upcoming 802.11g standard.

frequency hopping
A type of spread-spectrum signal that switches the transmission of data between frequencies based on a timing pattern that controls when the switch should occur.

GHz
Gigahertz; one billion hertz.

direct-sequence modulation
A type of spread-spectrum signal that transmits chips that can be reformatted by the receiver at different frequencies.

chips
The chunks of data that are transmitted within a direct-sequence modulation transmission.

orthogonal frequency division multiplexing (OFDM)
A type of radio signal used in wireless networks that divides the data signal across 48 separate sub-carriers. This leads to high data transmissions and superior performance.

Chapter 8

The following table shows the frequency ranges, maximum transmission distances, and bandwidths of the common wireless transmissions:

Type of Signal	Frequency Ranges	Maximum Transmission Distances	Bandwidth
Narrow-band single frequency	902–928MHz 2.4GHz 5.72–5.85GHz	50–70 meters	1–10Mbps
High-powered single frequency	902–928MHz 2.4GHz 5.72–5.85GHz	Line-of-sight	1–10Mbps
Frequency-hopping spread spectrum	902–928 MHz 2.4GHz	Can be several miles, depending on the limits of the cell	1–2Mbps
Direct-sequence-spread spectrum	902–928MHz 2.4GHz	Can be several miles, depending on the limits of the cell	2–6Mbps
Orthogonal frequency division multiplexing spread spectrum	5.15–5.25GHz 5.25–5.35GhH 5.725–5.825GHz	Up to several miles	Up to 54Mbps

Microwave Transmissions

Microwave systems—which transmit at specified speeds between 1GHz and 1THz—require an FCC license. There are two types of microwave transmissions that are typically used within MANs and WANs: terrestrial and satellite. (Some LANs do implement microwave transmissions, but it is very rare.) Microwave systems can cover much greater distances than radio transmissions—while radio is limited to several miles, microwave transmissions that use satellites can cover a global range.

TIP

All microwave transmissions are susceptible to weather conditions such as lightning.

Network Media

Terrestrial Microwave

Terrestrial microwave transmissions use the line-of-sight method for transmitting data between two antennas. These antennas are typically shaped like the dishes shown in the illustration below. These transmissions happen at much higher frequencies than radio transmissions—4–6GHz and 21–23GHz—and require a license from the FCC. Terrestrial microwave communication can take place over distances from 1 to 50 miles, provided that there is a direct line of sight.

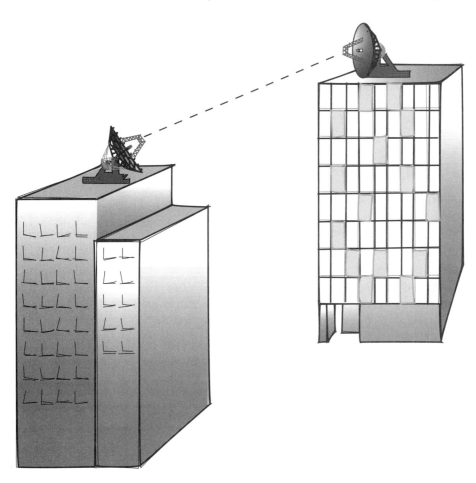

Chapter 8

Satellite Microwave

Satellite transmissions use the satellites orbiting the earth to pass on communications to the destination. The use of satellites allows transmissions to cover long, virtually unlimited distances. As you can imagine, launching a satellite to extend a network is extremely cost-prohibitive. Most implementations of satellite microwave network communications happen by paying a company to transmit the data. This is usually based on a per-minute fee and is extremely expensive.

The communication speed for satellite communications ranges between 1Mbps and 10Mbps. This type of communication is also subject to delays because of the distances that the signals must travel. Satellites use onboard **transponders** to transmit to the stations below. Although the transmission area of the transponders can be huge—even more than 10,000 kilometers across—they must still be in the line of sight of the stations.

transponder
A device for receiving radio signals that then converts the signals and transmits them in a different format.

infrared
Electromagnetic waves that can be used to transmit and receive wireless data transmissions.

Infrared Transmissions

Although **infrared** is used every day for transmitting signals from our remote controls to our television sets, it is not very practical in the networking environment. Infrared transmissions can be sent and received in one direction at a time or in both directions simultaneously. Communications that occur in one direction only can travel at up to 16Mbps, but communications that travel in many directions only reach a transmission rate of 1Mbps. The distance limitation for this type of transmission is 30 meters. These transmissions are also susceptible to interference from light sources, such as the sun or fluorescent lighting.

NOTE
Infrared transmissions use a frequency just below that of visible light.

NOTE
Recently, the FCC has approved infrared transmission in the terahertz range. Terahertz operates 1,000 times faster then gigahertz, or one trillion hertz per second.

Network Media

Comparing Network Media

The following table compares the network media discussed in this chapter, including the maximum segment length, bandwidth, and advantages and disadvantages of each type:

Media Type	Maximum Segment Length	Bandwidth	Advantages	Disadvantages
Thicknet coaxial	500 meters	10bps	Less susceptible to EMI interference than other types of copper media	Difficult to work with; expensive
Thinnet coaxial	185 meters	10Mbps	Less expensive than Thicknet or fiber-optic cable; easy to install	Limited bandwidth; limited application; damage to cable can bring entire network down
STP	100 meters	10Mbps	Reduced cross talk; more resistant to EMI than Thinnet or UTP	Difficult to work with; more expensive than UTP
Cat 3 UTP	100 meters	10Mbps	Least expensive of all media	Limited bandwidth; used primarily for voice applications; low signal quality
Cat 5 UTP	100 meters	100Mbps	Slightly higher cost than Cat 3 UTP but exceptional performance; easy to install; widely available and used	Susceptible to interference; can only cover a limited distance
Enhanced Cat 5 UTP	100 meters	1000Mbp s	Not much more expensive than Cat 5; greater bandwidth than Cat 5; improved cancellation	More expensive than Cat 5; more difficult to install than Cat 5; requires similarly rated connection devices

Media Type	Maximum Segment Length	Bandwidth	Advantages	Disadvantages
Single-mode fiber	3 kilometers	100Mbps–100Gbps	Excellent bandwidth compared to copper media; cannot be tapped, so security is better; not susceptible to EMI	Expensive; hard to terminate; can support only single transmissions
Multimode fiber	2 kilometers	100Mbps–9.92 Gbps	Can support multiple transmissions at a time; cannot be tapped, so security is better; can be used over great distances; not susceptible to EMI	Expensive; difficult to terminate
Single-frequency radio	50–70 meters; line of sight	1–10Mbps	Does not require installation of media	Can cover only a limited distance
Spread-spectrum radio	Up to several miles	1–54Mbps	Does not require installation of media	Can be expensive
Terrestrial microwave	1–50 miles	1–10Mbps	Does not require installation of media; can be used to connect to sites miles away	Less expensive than fiber-optic cable for some installations; susceptible to atmospheric conditions
Satellite microwave	Global	1–10Mbps	Does not require installation of media; available nearly anywhere in the world	Extremely expensive; susceptible to atmospheric conditions

Review Questions

1. What are the three types of network transmission media?

   _____

   _____

2. What are the three types of copper media?

   _____

   _____

3. What are two of the disadvantages of Thicknet coaxial cable?

   _____

   _____

4. What are two of the advantages of using Thinnet coaxial cable?

   _____

   _____

5. The distance limitation for a Thinnet segment is _____.

6. The connectors used with Thinnet are _____.

7. Where is STP typically found?

   _____

   _____

Terms to Know
- baseband
- cancellation
- chips
- cladding
- coaxial cable
- cross talk
- dielectric
- direct-sequence modulation (DS, DSSS)
- drop cable
- EMI (electromagnetic interference)
- fiber-optic cable
- frequency hopping (FH, FHSS)
- gauge

Review Questions

Terms to Know
- GHz
- gigabit
- IBM data connectors
- impedance
- infrared
- laser
- light-emitting diode (LED)
- media
- microns
- multimode
- OFDM
- ohms
- orthogonal frequency division multiplexing (OFDM)

8. What are two advantages of STP?

9. What is the most common unshielded twisted-pair cable used today?

10. What are two disadvantages of Cat 5 UTP?

11. How many pairs of wires are found in Category 5e UTP?

12. What is the transmission rate for Category 6 UTP?

13. What are the connectors used for UTP called?

14. Name the two types of fiber-optic cable.

_____

_____

15. What is the distance limitation for multimode fiber?

_____

_____

16. What are three types of wireless transmissions?

_____

_____

17. What are the advantages of wireless transmissions?

_____

_____

18. What are the differences between terrestrial and satellite microwave transmissions?

_____

_____

Terms to Know
- plenum
- RFI
- SC connector
- shielded twisted pair (STP)
- single-mode
- SMA connector
- ST connector
- Thicknet
- Thinnet
- transponders
- Underwriters Laboratories (UL)
- unshielded twisted pair (UTP)
- vampire tap

Chapter 9

Devices

Networking devices are fundamental to the interconnection and communication that takes place within a network. While some devices are specific to LANs, others can function at both the LAN and WAN level. As technology advances, more and more devices provide services at multiple layers of the OSI model. In this chapter, you will be learning about the types of devices you can use to connect and expand your network and the OSI layers where those devices function.

This chapter will cover the following topics that relate to networking devices used in Ethernet networks:

   How and why we extend our networks

   What network segments and collision domains are

   The types of devices found on a network

   The advantages and disadvantages of each device

Chapter 9

Extending the Network

When you build a simple network, it will include several computers and other resources, such as a printer and file server. In the following illustration, you'll see a simple network within the light gray circle.

access point
A device that connects wireless users to a network.

As additional resources are needed, you'll add other devices to extend the network. In the extended circle, you'll see additional repeaters and bridges that are used to connect LANs to additional resources. Wireless devices, such as **access points** and **wireless bridges**, allow the network to be extended without installing wires.

wireless bridge
A device that connects wired LAN segments using wireless transmissions.

The final connections added to the network include the router that allows you to communicate within a private WAN by again adding available resources. The router also allows you to connect to the Internet.

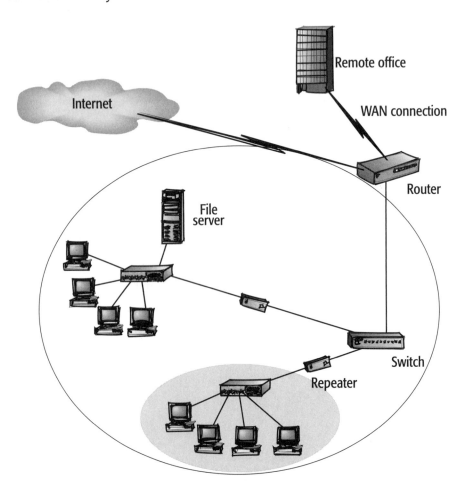

Network Segments

The term *segment* is used loosely among networking professionals. It's most commonly used when describing the cable used to connect two devices. The term **network segment** is more generally used when referring to a specific area of a LAN.

When there is a need to add devices to a network or extend the distance the network covers, performance can be seriously impacted. One of the disadvantages of extending your network to accommodate more devices is that the addition of devices increases network traffic. This problem can be remedied by other devices that filter traffic. These devices can reduce congestion by making sure only necessary information is passed along the network. One large network with many devices on it can be broken up into multiple network segments, which helps control traffic congestion and therefore improves performance. Each device on each network can still communicate with the rest of the network, but only when needed. The majority of the network traffic stays locally on the network segment. The size of a network segment is determined by two factors: broadcast domains and collision domains.

To understand broadcast domains, you need to know about broadcasts. A **broadcast** is a signal that is sent out from one device and read by all other devices attached to the same network. On all types of networks, broadcast messages can be used to find the names of all the computers on the network or to send a network-wide message. The problem with broadcasts is that they can unnecessarily consume network bandwidth, and this only gets worse as more devices are attached to the network. It's even a problem when no one is apparently using the network—the devices just continue to send out the broadcast messages at regular intervals.

A **broadcast domain** is defined by the boundaries that hold in the broadcasts. For example, if a broadcast message is sent down a cable and stops when it reaches a device, the message has reached the end of the broadcast domain. If the message is passed on by the device (as is the case with a hub), then the broadcast domain continues on until it reaches an end device or until it is blocked. There are some devices that can block broadcasts, but will allow other messages to get through. The messages that are allowed to pass are said to be entering a different network segment.

Another characteristic that defines the network segment is the **collision domain**. (You'll learn more about collisions in Chapter 10.) A collision happens when two devices try to transmit at the same time, causing the electrical charge of the signal to increase. When a collision occurs, all devices in the same collision domain sense the error and automatically back off for a period of time. In early Ethernet

---

network segment
The area of the network, bound by bridges or switches, in which collisions are propagated; or the area bound by a router to prevent the propagation of broadcasts.

broadcast
Data that is sent out to all devices on the network.

broadcast domain
All devices on a network that can receive the same broadcast packet.

collision domain
An area of the network, bound by bridges, switches, or routers, in which collisions are propagated.

networks, collision and broadcast domains covered the same area. The introduction of bridges and (more recently) switches has changed this distinction, as you will learn in the next section on networking devices.

Segmenting the LAN

Certain devices—such as bridges, switches, and routers—can be used to segment a LAN based on collision domains. Each port on a bridge or switch is a separate collision domain. These devices are designed to prevent collisions from being passed to other connected segments. Other devices pass collisions on to all the devices connected to them—repeaters and hubs propagate collisions and extend the collision domain. When using these devices, you must be cautious not to exceed the allowable number of nodes per network segment. This number is far greater than what you would ever hopefully attach to one network segment. (Technically, the number is 1024 nodes, but 30 nodes are considered a lot for one collision domain.)

The following illustration shows the multiple segments of a LAN:

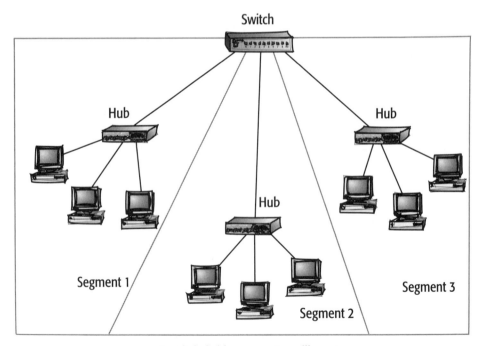

A switch, bridge, or router will create network segments. Hubs will not.

Network Interface Cards

The network interface card, or NIC, as it is commonly referred to, allows devices to connect to the network. NICs can provide connections for any type of networking media, including wireless media. Not only do servers and workstations have NICs, so do network printers and other network devices. A NIC uses a piece of software, known as a **driver**, to communicate with the computer's operating system. Without a driver, the NIC cannot function. Each Ethernet and Token Ring NIC has a physical address that is hard-coded into the device by the manufacturer. This is the Media Access Control address, or MAC address, a hexadecimal number that identifies the manufacturer that made the device. The MAC address must be unique so that data packets can be addressed appropriately to reach specific devices. If the NIC is replaced with another NIC, the MAC address also changes.

When one network device wants to communicate with another device, it looks in its **ARP table**. The ARP table contains the MAC addresses of other devices on the network. If you change hardware or add new hardware to your network, it may take a short period of time before the ARP table is updated. On some devices, such as routers, it may be necessary to clear the ARP table that is kept in memory. This flushes out any old information and forces the device to create an ARP table from scratch.

driver
Software used to interface between a device and the operating system.

ARP table
A list of MAC addresses mapped to the corresponding IP addresses of the workstations that they represent. The ARP table is maintained in the memory of the NIC.

motherboard
The main board in a computer that manages the communication and function of all information.

> **NOTE**
> Some NICs allow you to change the MAC address. This feature is rare and typically found only in some older cards.

There are three things you must look for when selecting a NIC for your workstation, laptop, or server:

- The NIC must support the type of network you wish to connect to. For example, if you are connecting to an Ethernet network, you must have an Ethernet NIC.
- The NIC must offer the right type of connector for the network media you're using—for example, a fiber-optic SC connector, a coaxial BNC connector, or an RJ-45 connector.
- The NIC must support the type of expansion slots you have within your **motherboard**, such as PCI or ISA slots.

Chapter 9

NOTE

Most laptops require a PCMCIA card that provides the appropriate type of network connection, including wireless.

Source and Destination

The NIC must also provide the other communication systems that allow the device to communicate on the network. The NIC functions at the Data Link and Physical layer of the OSI model, and provides many functions as both a source and a destination.

As a source device, the NIC:

- Receives the data packet from the Network layer
- Attaches its MAC address to the data packet
- Attaches the MAC address of the destination device to the data packet
- Converts data into packets appropriate to the network access method (Ethernet, Token Ring, FDDI)

Devices

- Converts packets into electrical, light, or radio signals to transmit over the network
- Provides the physical connection to the media

As a destination device, the NIC:

- Provides the physical connection to the media
- Translates the electrical, light, or radio signals into data
- Reads the destination MAC address to see whether it matches the device's own address
- Passes the packet to the Network layer if the destination MAC matches its own

Chapter 9

Repeaters

Repeaters are networking devices that allow you to connect segments, thus extending your network beyond the maximum length of your cable segment.

repeater
A network device that regenerates and propagates signals between two network segments.

A repeater functions at the Physical layer of the OSI model and connects two segments of the same network. A multiport repeater, known as a *hub*, connects several segments together. A repeater can connect segments of the same network that use different media types, such as an Ethernet segment using coaxial cable and another using Cat 5.

A repeater has three basic functions:

- It receives the incoming signal and cleans it up.
- It re-times the signal to avoid collisions on the network.
- It transmits the signal onto the connected segment.

The following illustration shows how a repeater can extend beyond the limits of a single cable segment the distance a network can reach:

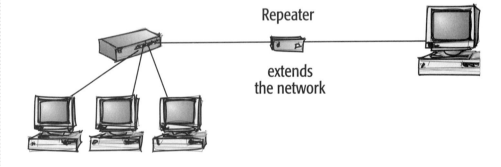

TIP
Using a repeater can help regenerate signals as they travel through areas with high levels of EMI (electromagnetic interference).

Devices

Hubs

A **hub** is the central point of connection for cable segments in a physical star topology. It allows the connection of multiple devices to the network and can provide different services, depending on the sophistication of the hub. You may see advertisements for managed hubs, switched hubs, and even "intelligent hubs," each of which provides additional services beyond what a basic hub provides. You will need to know what services you want so that you can select your hub accordingly.

While the term *hub* refers to a specific networking device, in reality a hub is a repeater with many **ports**. As a multiport repeater, a hub functions in the same way as a repeater, although in addition to the functions of the repeater, a hub must regenerate the signal for all connected segments.

Hubs work at the Physical layer of the OSI model. They pass along all data that they receive, no matter which device it is addressed to, which can add to congestion on the network. The following illustration shows a simple four-port hub connected to networking devices and creating a physical star topology:

hub
A LAN device that broadcasts the information it receives to all attached devices and segments of the network.

port
The female interface on an internetworking device. Used with jacks to form a connection.

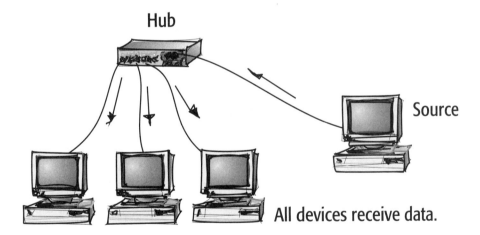

WARNING

As you can imagine, using a large hub (such as a 24-port hub) or stacking hubs together can severely impact the network's performance because all the devices must listen to all the other devices and wait to transmit until all the other traffic has stopped.

189

Chapter 9

When choosing a hub for your Ethernet network, you will need to consider the following:

The type of media connection you'll need.   Typically, hubs provide one type of network connection, although some do provide another port of a different type. Many 10BaseT hubs provide one BNC port with the other RJ-45 ports.

The number of ports you'll need.   Hubs are available in 4- to 48-port configurations. It's now possible to stack together 24-port hubs for as many as 96 ports.

Speed.   Typical hubs transmit and receive at 10Mbps. Some hubs have one or more ports that can transmit and receive at either 10Mbps or 100Mbps, or are capable of switching between the two speeds, depending on the device that is attached.

Whether it's managed or unmanaged.   Managed hubs allow a network administrator to use software to view how the device is functioning from a remote workstation. Unmanaged hubs don't.

Whether there's an uplink port.   The **uplink port** is necessary when you interconnect two hubs using a patch cable. Many hubs have a small switch located next to the uplink port that can be used to make the port an uplink port or a regular port that can be used to attach end devices. Without an uplink port, you would need to use a **crossover cable** between the hubs. The switch or the crossover cable (but not both together) can be used to allow the hubs to transmit to each other.

uplink port
The port on a hub or switch that is used to extend the network to another hub, repeater, or switch.

crossover cable
A special cable that reverses the transmit and receive wires from one end to the other to directly interconnect devices without a hub.

> **NOTE**
> Token Ring hubs, which are called MAUs, provide the same central connection for devices in a physical star. MAUs function differently from Ethernet hubs in that the data travels from device to device in a logical ring. See Chapter 5 for more about physical and logical topologies.

Devices

Access Points

Access points, called APs, are hardware devices or software that act as a wireless hub. When used in conjunction with a traditional wired LAN, an AP allows wireless users to connect to the wired network's services. When an AP is used to connect to a wired network, the wireless network is said to be in **infrastructure mode**. The following image shows an AP connected to a network switch providing wireless services to the wireless devices.

infrastructure mode
A type of wireless LAN that utilizes an access point through which wireless devices communicate.

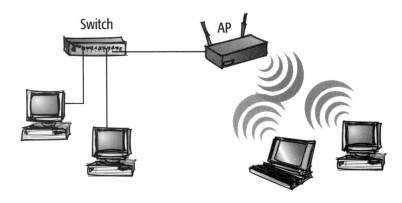

ad-hoc mode
A type of wireless LAN that does not utilize an access point; devices communicate directly with each other.

Wireless Equivalent Privacy
An encryption method used in wireless networks; typically called WEP.

See Chapter 5 for more on infrastructure and **ad-hoc modes** in wireless networking.

In home networks, wireless access points are commonly added to a DSL line, cable modem, or even a 56K dial-up connection. In this way, multiple users can access the Internet or other devices using one device.

APs can also be used independently, without connecting to a wired LAN. In this way, they extend the range of wireless communication within the WLAN; several wireless devices can communicate with each other via the AP. APs can support multiple users, depending on the capacity of the device.

Access points, like all wireless transmissions are not as secure as wired network devices. Potentially, anyone with a wireless NIC could access services from your access point. On many devices, you can select the level of security you would like. A low level of encryption would be 64-bit; a higher level would be 128-bit.

> **NOTE**
>
> **Wireless Equivalent Privacy, WEP** is a method of encryption used in 802.11b wireless networks.

See Chapter 10 for more on WEP.

191

Chapter 9

Bridges

bridge
A LAN device that will filter or forward data packets based on the MAC address of the destination. It works between network segments that use the same protocol.

Bridges, like hubs, connect LAN segments, but they work at the Data Link layer of the OSI model. Because bridges work at the Data Link layer, they can use the MAC addresses to make decisions about the data packets they receive. A bridge provides four key functions:

- ○ It builds a bridging table to keep track of devices on each segment.
- ○ It filters packets that do not need to be forwarded to other segments based on their MAC address.
- ○ It forwards packets whose destination MAC address is on a different network segment from the source.
- ○ It divides one network into multiple collision domains, thereby reducing the number of collisions on any network segment.

Bridges use the Spanning Tree Protocol (STP) to decide whether to forward a packet through the bridge and on to a different network segment. STP serves two functions: One is to determine a main bridge, called a *root*, which will make all the bridging decisions and deal with all bridging problems. The second function is to prevent bridging loops.

The following illustration shows how a bridge receives data, filters it based on the MAC address, and forwards it if necessary:

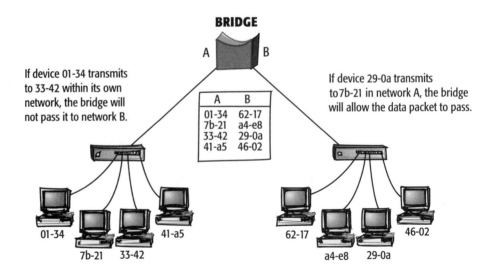

192

Wireless Bridges

Wireless bridges are designed to connect two or more networks, often in different buildings. Wireless bridges are much more cost-effective than running fiber-optic cabling or leasing lines from a provider. Wireless bridges can provide 10Mbps or even 100Mbps of bandwidth, depending on the type you purchase. Naturally, the greater the bandwidth, the larger the price tag.

Wireless bridges can also be used to connect wireless clients to the network, much like an access point. Wireless bridges that also act as access points can typically cost two to three times more than a typical access point. The following figure shows a wireless bridge that connects two LAN segments and also provides wireless access to clients.

Chapter 9

Switches

switch
A LAN device that filters or forwards data packets to the correct port on a device based on the MAC address of the destination.

Switches are actually multiport bridges that function at the Data Link layer of the OSI model. There are Ethernet, Token Ring, FDDI, ATM, and many other types of switches (including some that function at higher layers of the OSI model). Each port of a switch makes a decision to forward data packets to the attached network. The switch keeps track of all attached nodes' MAC addresses and which port on the switch each node is attached to. This information can help filter traffic and eliminate unwanted congestion, just like with a bridge.

Switches have become increasingly popular with network administrators and designers. Because the prices of switches have dropped to a level that makes them competitive with hubs, many of today's networks have switches installed in their wiring closets. Switches cut down on network traffic and keep the transmission of bandwidth-intensive data from affecting the entire network. Like a bridge, each port on a switch is a separate collision domain. If only one device is attached to a port on a switch, there is little chance that there will be a collision. The following illustration shows how a switch connecting segments of a network filters traffic that would otherwise be passed along to all devices by a hub:

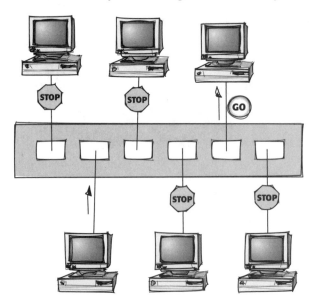

Virtual LANs

Switches have become extremely sophisticated in the last few years. In addition to supporting fiber-optic networks and extremely high speeds, switches have

Devices

added services at the upper layers of the OSI model. At the Application layer, many switches have a built-in web server for management; this feature comes on most of Cisco's Catalyst switches, and is very useful. The most important advancement in switching that we've seen is the capability to create **virtual LANs (VLANs)**.

VLANs are used to create broadcast domains on the same device. You learned earlier that ports on bridges and switches are each individual collision domains. With VLANs, ports can be grouped into a single broadcast domain. This type of VLAN can be created at layer 2 using MAC addresses or at layer 3 using network addresses (like IP). VLANs have their own standard (IEEE 802.1P and 802.1Q) that specifies the use of a router to pass packets between VLANs and to other networks. The following illustration shows how a VLAN allows devices to function as part of the same network segment, despite their physical location and connection:

virtual LAN (VLAN)
A LAN in which devices are logically configured to communicate as if they were attached to the same network, without regard to their physical locations.

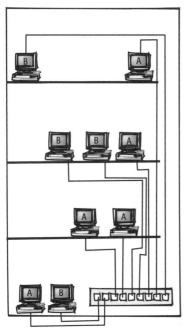

Devices with the same letter are a part of the same network. They cannot see the other network.

TIP
Switches can support one or many MAC addresses per port. Be sure to check whether the device you purchase will meet your needs. For example, a switch that supports only a single MAC address per port can't have a hub directly attached to it because the hub has the potential of attaching many devices that each has its own MAC address.

Brouters

brouter
A hybrid device that bridges non-routable layer 2 protocols, and routes routable layer 3 protocols.

The name **brouter** comes from the terms *bridge* and *router*. A brouter is a hybrid device that can function as both a bridge and a router, and can be used to interconnect multiple networks. A brouter can work on networks that are using many different protocols, and allows a network to exceed the maximum number of devices allowed.

Brouters, like more advanced switches, can be programmed to function in a specific way. If a brouter is programmed to forward data packets that are of a type of protocol that is non-routable, such as Microsoft's NetBEUI, the brouter uses the layer 2 MAC address to decide whether to forward the packet or not. In this way, it is functioning as a bridge. NetBEUI is a non-routable layer 2 protocol.

If a brouter is set to route data packets to the appropriate network with a routed protocol such as IP, it is functioning as a router. As a router, it is working at the Network layer. IP is a routable layer 3 protocol. The following illustration shows a brouter functioning as both a bridge and a router:

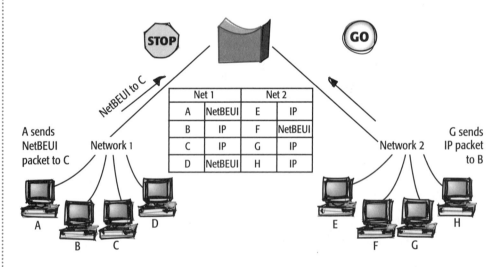

Brouters—bridge layer 2 protocols and route layer 3 protocols

Devices

Routers

Routers are Network layer devices that are extremely intelligent. Routers can connect different network segments that may be located in the same building or thousands of miles apart. They work in LAN, MAN, and WAN environments. Routers allow you, the user, to access resources within or beyond your LAN by using the best path to the resources. They use the network address—the IP address in the TCP/IP environment—to determine the best path for the packet to take to reach the destination quickly.

Routers can also be used to interconnect different types of networks, such as Token Ring and Ethernet networks. The router changes the packet's size and format to fit the type of destination network on which the packet is being sent. In addition, routers connect networks that use one or many of the following **routed protocols**: Internet Protocol (IP), Internetwork Packet eXchange (IPX), DECnet, and AppleTalk.

> **TIP**
>
> Not all routers can accommodate all types of networks. The type of network a router can connect to is determined by both the router hardware and software.

How Routers Select the Best Path

A router has two primary functions: to determine the best path to a destination and to share path information (the "route") with other routers. A router determines the path data should take—based on one of two types of routing: **static routing** or **dynamic routing**. Whichever type of routing is used, the router determines the path for a packet by looking at the **routing table**, in which information about other networks is stored by the router. The table keeps track of known networks, the port on the router that should be used to send data to a particular network, and the "cost" for a data packet to get to that network. The cost is based on an algorithm determined by the **routing protocol** being used, or it is set manually by the network administrator.

> **NOTE**
>
> A letter and a number identify router ports. On Cisco routers, the serial port—the port used to connect to the telco—is identified by an S; the Ethernet port is identified by an E. If there are multiple ports of each type, the first port on the router always uses the number 0.

---

router
A LAN and WAN device that determines the best path to send network traffic based on cost and Network layer information.

routed protocol
A LAN protocol that can be transmitted to other networks by way of a router; examples include AppleTalk, DECnet, IP, and IPX.

static routing
A type of routing in which the network administrator manually configures a route into the router's routing table.

dynamic routing
Internetwork routing that adjusts automatically to network topology or traffic changes based on information it receives from other routers.

routing table
A table that keeps track of the routes to networks and the costs associated with those routes.

routing protocol
A protocol that uses a specific algorithm to route data across a network.

The following illustration shows a basic router and the ports used to connect it to the network. While some routers only have two ports, others can have a high number of ports to accommodate greater network connections.

The next illustration shows several networks connected with routers. Even though there are several paths that the data could take to travel from network 1 to network 6, the router will determine the best path, based on the cost associated with getting there.

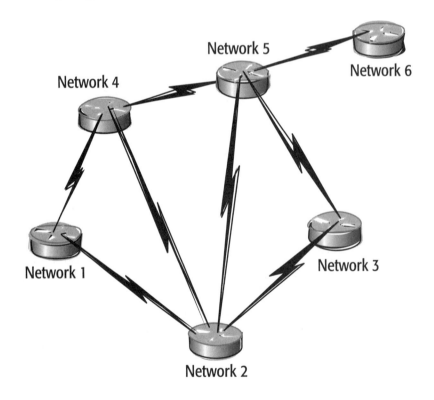

Devices

Gateways

Gateways are devices that allow different types of network systems to communicate. They provide translation services to systems that could otherwise not communicate. Typically, gateways are nodes on the network that provide these services. Depending on the type of translation they are providing, they work at every level of the OSI model. They are used for a variety of networking tasks, and can be categorized by the services they provide. There are three main types of gateways:

gateway
A device that performs the translation of information from one protocol stack to another.

Address gateway   An address gateway connects networks using the same protocol, but using different directory spaces such as MHS (Message Handling Service).

Protocol gateway   A protocol gateway connects networks that are using different protocols. The gateway translates the source protocol so that the destination network can understand it.

Application gateway   When a standards-based Internet e-mail server needs to send mail to a messaging server, a gateway is used to translate the Internet e-mail into a format that the messaging server can understand.

NOTE

The term *gateway* is used in a very general way in networking, and can mean a variety of things. If the term is used in reference to a device on the network, it means the device is providing one of these types of translation services. A router is often called a *gateway* because it performs translation services between different networks at the Physical, Data Link, and Network layers.

Comparing Networking Devices

The following table compares the networking devices discussed in this chapter:

Device	Advantages	Disadvantages
Repeater	Can connect different types of media; can extend the distance a network can reach; does not increase network traffic	Extends the collision domain; cannot filter information; cannot connect different network architectures; only a limited number can be used within a network
Hub	Cheap; can connect segments using different media types	Extends the collision domain; cannot filter information; passes along all packets to all connected segments
Access point	Relatively cheap for use with a small number of wireless devices; does not require a physical connection for users to access network resource	Reduced security; distance limitations; can be affected by interference or structures
Bridge	Limits the collision domain; can extend network distances; can filter packets based on their MAC addresses and ease congestion; can connect different types of media; some can connect different types of network architectures	Broadcast packets cannot be filtered; more expensive than a repeater; slower than a repeater due to processing the addresses and filtering packets
Wireless bridge	Allows the wireless connection of multiple wired LAN segments; can provide services to wireless clients	More expensive than an access point; if large amounts of throughput are needed, it can be expensive
Switch	Limits the collision domain; can provide bridging to multiple segments of a network; can be configured to limit broadcast domains through the use of virtual LANs	More expensive than a hub or a bridge; configuration of additional functions can be complex
Brouter	Limits the collision domain; can provide the services of bridges and routers in one device	More expensive than a bridge

Devices

Device	Advantages	Disadvantages
Router	Limits the collision domain; can function in the LAN or WAN environment; can connect networks using different media and architectures; can determine the best path for a packet to reach another network; can filter broadcasts	Expensive; must be used with routable protocols; must be configured by an administrator; can be difficult to configure; slower than a bridge due to increased processing and routing updates sent between routers
Gateway	Can connect different network systems using different protocols, addresses, and applications	Can be difficult to install and configure; additional overhead for processing data slows network performance

Review Questions

Terms to Know
- access point
- ARP table
- bridge
- broadcast
- broadcast domain
- brouter
- collision domain
- crossover cable
- driver
- dynamic routing
- gateway
- hub
- infrastructure mode
- motherboard

1. Which device is responsible for the connection between a device and the network media?

2. What is the MAC address?

3. At which layer of the OSI model does a NIC function?

4. Which networking devices function at the Physical layer?

5. How does a hub function?

6. What is a collision domain?

7. What is a network segment?

8. What is an access point?

9. How does a bridge function?

10. What is a wireless bridge?

11. What is a switch?

12. What is the function of a virtual LAN?

13. At which layer of the OSI model do bridges and switches function?

14. What is the purpose of a router?

15. What does a router use to make decisions about forwarding packets on to other networks?

16. What are the two types of routing?

17. How is a static route determined?

18. How is a dynamic route determined?

19. What is the function of a gateway?

20. At which layers of the OSI model do gateways function?

Terms to Know
- network segment
- port
- repeaters
- routed protocol
- router
- routing protocol
- routing table
- static routing
- switch
- uplink port
- virtual LAN (VLAN)
- wireless bridge
- Wireless Equivalent Privacy (WEP)

Chapter 10

Standards

Standards define almost every aspect of our daily lives, from which side of the road we drive on, to the type of electrical outlets we have in our homes, to how we do our jobs. In order for a standard to be effective, it must be agreed upon by the people using it. Imagine, for example, that a group of individuals invents a new type of electrical plug. The plug might work in their laboratory, but if no outlets can accept it in the real world, the plug is essentially useless. It will never be used by the general populace unless it is accepted as a standard.

National and international organizations work together to define the standards that organize our lives. Understanding how and by whom standards are adopted, as well as their importance in allowing the intercommunication of devices, is necessary to those who work in the networking field.

In this chapter, you will learn:

   How a standard is developed and adopted

   The major standards organizations

   The IEEE standards for communication

   WAN standards and access methods

What Are Standards?

standard
A set of rules or procedures that is officially accepted.

Request for Comment (RFC)
A public request for feedback about a standard being developed.

Standards are sets of rules that are agreed upon by a group. Each organization involved in adopting a standard is made up of a group of experts in their field. Standards define everything from safety requirements for things like bicycle helmets to quality control in the manufacturing process. In networking and internetworking, standards are adopted by a relatively small number of organizations. Manufacturers may play a part in the adoption process, but typically manufacturers follow standards to ensure that their products function in the real world.

Standards in networking allow for the intercommunication of devices that may be very different and located within different networks. By using standards that define the rules for communication, all devices can communicate. Likewise, by following standards, a network administrator can implement the most secure network possible. Following the rules for the number of devices within a network segment and the maximum segment length prevents unnecessary problems on the network.

The adoption process is not an easy one. Different organizations approach the process in different ways. Most networking standards are adopted either through the public Request for Comments (RFC) process or by consensus/voting done within the major standards organizations. Some organizations, such as the American National Standards Institute (ANSI), provide an opportunity for qualified groups to reach consensus about a standard, rather than develop the standard themselves.

How Does an Idea Become a Standard?

Standards follow a life cycle similar to that of laws. Laws begin as ideas known as bills, which members of Congress and other interested parties discuss. Bills get changed and revised and occasionally ignored. Sometimes, enough people agree on the worth of a bill, and the bill eventually becomes a law. Similarly, an idea for a new networking standard begins as a **Request for Comments (RFC)**.

When an organization has developed its idea for a standard to the point of needing outside ideas about its work, the organization sends out an RFC to gather ideas and information about the standard that it is proposing. The RFC is posted to the Internet or an electronic bulletin board, where interested parties can look at it and provide input. You can find all the RFCs that have ever been published on many sites on the Internet. One resource is the RFC Archive at http://www.sunsite.auc.dk/RFC/rfc.

Standards

The First RFC: The Beginning of the Internet

In 1969, Steve Crocker at UCLA sent out the first RFC to all of the people working on the Internet project. The topic was host software. The RFC allowed interested members of the Network Working Group to see the work that was being done and to make additional comments that might result in changes and improvements.

This first RFC is very interesting because it outlines how the Internet and communications between remote devices were first envisioned. In this RFC, IMP stands for *interface message processors*, which are now referred to as *packet-switching nodes*. Packet switching is an instrumental part of transmitting data through the telephone network.

```
RFC 1:
Title:          Host Software
Author:         Steve Crocker
Installation:   UCLA
Date:           7 April 1969
Network Working Group Request for Comment: 1
CONTENTS
INTRODUCTION
    I. A Summary of the IMP Software
        Messages
        Links
        IMP Transmission and Error Checking
        Open Questions on the IMP Software
   II. Some Requirements Upon the Host-to-Host Software
        Simple Use
        Deep Use
        Error Checking
  III. The Host Software
        Establishment of a Connection
        High Volume Transmission
        A Summary of Primitives
        Error Checking
        Closer Interaction
        Open Questions
   IV. Initial Experiments
        Experiment One
        Experiment Two
```

Once an RFC is posted, people can make comments to the author. If there are changes that are decided upon based on these ideas or comments, a new RFC is posted with a new number. No RFC is changed after it has been posted. Currently, more than 2,500 RFCs have been posted for comment. Some of the more recent RFCs are related to security on the Internet.

Major Standards Organizations

There are six main organizations that adopt standards that relate to local, metropolitan, and wide area networks. They work to provide interoperability between devices, define how we design and implement our networks, and provide a basis for the ways our networks communicate. The main organizations you should be aware of are:

- International Organization for Standardization (ISO)
- Institute of Electrical and Electronics Engineers (IEEE)
- Electronics Industries Alliance (EIA)
- Telecommunications Industry Association (TIA)
- American National Standards Institute (ANSI)
- International Telecommunications Union (ITU)

Even though these organizations are located in different countries and continents, their standards are usually adopted worldwide. There are times, though, when one group will adopt a standard, and another will adopt one that's similar, though not exactly the same. This can lead to confusion within the related industry. Knowing which standard to implement can be challenging until they are adopted internationally.

Members of the organizations listed previously participate in the development of standards. Often, members of these groups are electrical, civil, and mechanical engineers; computer science engineers; electricians; telecommunications companies; scientists; physicists; and other professionals involved in the telecommunications industry. Some of the standards organizations allow any individual who is working in their field to become a part of the process.

Not all of these organizations work on the same telecommunications standards issues, nor do all implement or develop international standards. For example, ANSI is responsible for standards for the United States, whereas the ITU oversees international standards.

There are many different national and international organizations that deal with standardization issues specific to their industry. There are also many organizations and businesses that work within the networking field. These six major standards organizations have adopted the majority of the internetworking standards we use.

Standards

ISO: International Organization for Standardization

The **International Organization for Standardization (ISO)**—www.iso.ch—is the granddaddy of standards organizations. Established in 1947, the ISO is located in Geneva, Switzerland, and is actually a collection of other standards organizations from around the world. There are more than 130 countries represented in this non-government group. Together, the members of the ISO have been responsible for adopting one of the most important standards in networking and internetworking: the OSI reference model.

Because the ISO is an international organization, the standards that it adopts are international standards. The ISO's impact goes far beyond its adopting the OSI model. Some of the standards it has adopted—and you have benefited from—are film speed codes, telephone and banking cards, and even the type of container used for freight. To date, the ISO has adopted more than 12,000 international standards.

International Organization for Standardization (ISO)
An international organization that develops standards, including those relevant to networking. The ISO developed the OSI model used as the theoretical model for intercommunication between networking devices.

ISO's Three-Step Process

When the ISO adopts a standard, it follows a formalized three-step process, which is very different from the posting of RFCs. This process can take a long time. As a result, the standard may have competition from a standard adopted by a smaller group.

Step 1: Defining the Scope   The adoption process begins when a group or industry expresses a need for a standard. The industry discusses whether there is a valid need for the standard. Once all parties agree that there is, the ISO defines what the impact and scope of the standard will be after it is adopted. This work usually takes place in committees.

Step 2: Negotiations   The second phase is the negotiations. All the countries represented in the ISO have an opportunity to give input as to the standard's final form. In the end of this negotiation process, an agreement about the proposed standard is made through consensus. Without consensus, the standard will not be adopted.

Step 3: Formal Adoption   The formal adoption of the standard has two components: First, the members of the committees that actually worked on the proposed standard must pass the draft of the standard with two-thirds agreement. The second approval must come from all of the voting members of the ISO, who must pass the proposal with 75 percent in agreement. After these votes are taken, the proposal is published as an ISO international standard.

Chapter 10

IEEE: Institute of Electrical and Electronics Engineers

Institute of Electrical and Electronics Engineers (IEEE)
An organization that develops and adopts communications and networking standards, including the 802 series of LAN standards.

The **Institute of Electrical and Electronics Engineers (IEEE)**, commonly pronounced "eye-triple-ee," is a group of professionals from the scientific, technical, and educational fields. Members can join the IEEE—and yes, even you can join—by registering at its website (www.ieee.org). The IEEE members, like those of the ISO, represent many countries throughout the world—more than 140.

The most common standards adopted by the IEEE are the 802 standards, which define physical cabling requirements and the way data is transmitted. While one standard may be adopted and given a number, changes to, additions to, or special applications of the standard may be defined using the same 802 number with an additional letter. For example, 802.3u is the Fast Ethernet standard.

Electronics Industries Alliance (EIA)
A group of professionals that develops electrical transmission standards. The EIA and TIA work jointly to develop many communications standards.

NOTE

All of the IEEE standards are outlined at the website standards.ieee.org. You can search IEEE's database for information and abstracts on all the standards and drafts it has developed.

EIA/TIA: Electronics Industry Alliance and Telecommunications Industry Association

EIA/TIA standards are a combined effort between the EIA (Electronics Industry Alliance) and the TIA (Telecommunications Industry Association). EIA is often identified as the Electronics Industry Association.

EIA: Electronics Industry Alliance

The **Electronics Industry Alliance (EIA)** is a group of organizations that work together to develop and improve electronic devices, connections, and standards. The groups that make up the alliance include the Consumer Electronics Manufacturers Association (CEMA); the Electronic Components, Assemblies, Equipment, and Supplies Association (ECA); the Telecommunications Industry Association (TIA); the Electronic Information Group (EIG); the Government Electronics and Information Technology Association (GEIA); the JEDEC Solid State Technology Association; and the Electronics Industries Foundation (EIF).

Standards

TIA: Telecommunications Industry Association

The **Telecommunications Industry Association (TIA)** is a national trade organization made up of large and small companies that work in the telecommunications industry. The members of TIA manufacture the majority of products that are used in today's networks. The standards that they develop are acknowledged by ANSI and are voluntary industry standards—products are not required to meet them. Their standards address a variety of telecommunications products.

The standards adopted by the EIA/TIA have a great impact on the installation of a network. There are four main EIA/TIA standards that should be implemented when installing a network. The four most important standards to know are:

The 568 standard Defines the installation of networking media and termination.

The 569 standard Outlines the selection and requirements for wiring closets.

The 606 standard Defines the labels found on media.

The 607 standard Outlines the grounding requirements for buildings and wiring closets.

Remember that these standards are recommendations, so businesses and individuals are not required to follow them. Companies that do, however, will see that its network will have more flexibility for the future, and support a variety of topologies and speeds up to and beyond 100Mbps.

> **Telecommunications Industry Association (TIA)**
> An organization that works to develop and adopt telecommunications technologies standards. The TIA and EIA work jointly to develop many communications standards.

> **American National Standards Institute (ANSI)**
> An organization that coordinates the development of standards, including those in the area of communications and networking. It also approves U.S. national standards, and is a member of the IEC and the ISO.

NOTE

The Electronics Industry Alliance's web site is located at www.eia.org.

NOTE

The Telecommunications Industry Association's web address is www.tiaonline.org.

ANSI: American National Standards Institute

The **American National Standards Institute (ANSI)** is responsible for the development of standards relating to communications and networking. Standards for programming languages are an important part of this. ANSI works to establish consensus between qualified groups about standards; it does not develop standards itself.

ANSI also represents the United States as a member of the ISO.

International Telecommunications Union (ITU)
An international organization that develops communication standards.

> **NOTE**
> ANSI's web address is web.ansi.org.

ITU: International Telecommunications Union

The **International Telecommunications Union (ITU)** is a part of the United Nations, and works to develop international telecommunications standards. It includes three separate agencies:

- The Consultative Committee for International Telecommunications and Telegraphy (CCITT) is responsible for standards relating to communications. It is now called the ITU-T, the International Telecommunications Union-Telecommunications. The CCITT/ITU-T is responsible for many of the WAN standards in use today.
- The International Frequency Registration Board (IFRB) allocates telecommunications frequencies. This organization is now known as the ITU-R.
- The Consultative Committee on International Radio (CCIR), which is now part of the ITU-R, provides recommendations relating to radio communications.

> **NOTE**
> The International Telecommunications Union's web address is www.itu.int.

Standards

IEEE 802: Standards for Local and Metropolitan Area Networks

The 802 standard is an overview of the standards that define the Physical layer and MAC sublayer connections of LAN and MAN devices. All of these 802 standards, although different at the lowest layers of the OSI model, are compatible and work together at the Data Link layer.

There are currently fourteen 802 standards that are used to define LAN and MAN networks. Each of the standards focuses on one area of intercommunication at the Physical and/or Data Link layer of the OSI model.

TIP

The standards that are adopted by the IEEE are not openly available for public viewing; they must be purchased or subscribed to. Individuals and business must pay to acquire the information they need to develop a product that will meet the standard.

802.1: LAN and MAN Bridging and Management

The 802.1 standard defines the way in which a networking device, such as a bridge, selects a path to connect local area networks and metropolitan area networks. One of the standards that is defined under 802.1 is the **Spanning Tree Protocol (STP)**.

802.2: Logical Link Control

The 802.2 standard defines the upper portion of the Data Link layer, known as the Logical Link Control (LLC) sublayer, which uses the Logical Link Control protocol. The LLC protocol is responsible for providing connection-oriented service. (Connection-oriented and connectionless transmissions are discussed in Chapter 2.)

The LLC protocol uses an extended two-byte address. The first byte indicates a destination service access point (DSAP), and the second indicates a source service access point (SSAP). The LLC protocol also uses a **Subnetwork Access Protocol (SNAP)** header, which contains IP address packets.

Spanning Tree Protocol (STP)
A layer 2 protocol used in bridges and switches to identify the shortest path to any device on a local area network.

Subnetwork Access Protocol (SNAP)
An Internet protocol that specifies the encapsulation method for IP datagrams and ARP messages, and provides a link between a subnetwork device and an end system.

Chapter 10

802.3: CSMA/CD Access Method

Carrier Sense Multiple Access/Collision Detection (CSMA/CD) is commonly known as Ethernet, although technically Ethernet defines the cable and CSMA/CD defines the way that the cable is accessed. CSMA/CD requires that all devices on the network must listen to the cable before it can transmit. This is the **Carrier Sense** part of the standard.

Carrier Sense Multiple Access/Collision Detection (CSMA/CD)
The access method that is the basis for Ethernet and works at the MAC sublayer of the OSI Data Link layer.

Only one device on the network can transmit at a time. So if a device senses a transmission already on the wire, it must wait. Once the line is clear, the device can send its message. This is the **Multiple Access** part.

In the following illustration, you can see that devices are waiting because there is a message on the wire. Once the line is clear, one of the other devices may transmit.

backoff
After a collision, all devices on an Ethernet network wait a random amount of time before transmitting data.

Collision Detection

If, by some misunderstanding, more than one device sends a message at the same time, a collision occurs. This collision spreads throughout the segment in which it occurred. Thus, all the devices learn about the collision and must back off from transmitting. This is the *Collision Detection* part of the standard.

In the next illustration, two devices have transmitted on the line at the same time, resulting in a collision. The devices have stopped transmitting as a result of the **backoff** that was issued. No messages are being sent.

The backoff is the amount of time that a device has to wait before trying to send another message. The backoff period is based on a binary algorithm.

Standards

When a device has waited its allotted time, it then listens to the line again before it sends a message. This is often referred to as the *listen before transmit* method.

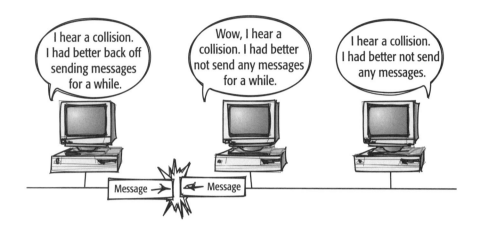

802.3u: Fast Ethernet

One of the standards that is an extension of the 802.3 standard is the 802.3u standard. It is most commonly known as **Fast Ethernet**.

The 802.3u standard was created out of the growing bandwidth needs of users. The original standard for 802.3 defined cabling types, distances, and the number of devices per segment based on a transmission speed of 10Mbps. These rules did not work when devices began transmitting at faster speeds.

The standard for Fast Ethernet created new specifications for cable construction, and set the maximum cable segment lengths for networks transmitting at 100Mbps. Under this standard, the cable that is installed must meet new requirements for handling faster transmission speeds, and the cable length requirements are shorter for each segment. Fast Ethernet continues to use the CSMA/CD access method. For this reason, it is an addendum to the 802.3 standard, not a new standard.

802.3z: Gigabit Ethernet

In July of 1998, the IEEE adopted the 802.3z standard. It utilized the 802.3 CSMA/CD access method and frame format with the addition of a carrier extension field. It provides transmission speeds of 1,000Mbps or 1Gbps. Because it is based on 802.3, devices that utilize the 802.3z standard can also communicate with devices transmitting at 10 or 100Mbps. The best thing about 802.3z is that it supports all existing network protocols at much higher speeds.

Fast Ethernet
A network technology defined by the IEEE 802.3u standard for 100Mbps networks.

Gigabit Ethernet
A network technology defined by the IEEE 802.3z standard for 1,000Mbps or 1Gbps networks.

gigabit
A unit of storage defined as 1,073,741,824 (approximately one billion) bits.

Chapter 10

802.4: Token Passing Bus Access Method

The 802.4 standard was developed to provide the benefits of Token Ring without the physical requirements of a ring. Physically, an 802.4 network is installed in a line or tree-shaped cable. Logically, the data actually travels through the wires in the form of a ring.

token
A frame passed on a Token Ring network. Possession of the frame allows a device to transmit data on the network.

cell
A fixed-size packet. In ATM networks, the cell size is fixed at 53 bytes.

802.5: Token Ring Access Method

Although technically Token Ring can use true physical ring cabling, the most commonly used topology is physical star, logical ring. Under the 802.5 standard, devices take turns transmitting data. They are able to transmit if they receive a **token** that is empty. The data they have to transmit is placed in the token and passed on to the next device in the network. In this way, the token travels around the entire network looking for the device it is addressed to.

When the destination device receives the token, it takes its message out and replaces it with a message for the source device that it received the message from. Once the source device gets the token back and takes out its own message, the next computer in the ring gets a chance to use the token to transmit.

> **NOTE**
> It's interesting to note that 802.5 came out after Token Ring. So the standard, while describing Token Ring, isn't really Token Ring (as is trademarked by IBM).

802.6: DQDB Access Method

The Distributed Queue Dual Bus (DQDB) standard was developed for metropolitan area networks. This standard uses two parallel cables in a bus network topology. The bus has a head that generates **cells** that travel throughout the network until they reach the end of the cable. Each of the two cables in the bus transmits in a different direction so devices can send messages either "upstream" or "downstream." (For more information about bus topologies, see Chapter 5.)

802.7: Broadband Local Area Networks

The 802.7 standard specifies the design, installation, and testing necessary for broadband transmissions, which allow for multiple transmissions using different

Standards

channels at the same time. The broadband bus topology creates a full-duplex medium that supports **multiplexing**.

802.8: Fiber-Optic Local and Metropolitan Area Networks

The 802.8 standard states the recommendations for configuring and testing fiber-optic LANs and MANs. The testing specified under this standard ensures the integrity of the fiber-optic cabling. (For information about fiber-optic cabling, see Chapter 8.)

802.9: Integrated Services

The 802.9 standard defines a unified access method that offers integrated services (IS) for both public and private backbone networks, such as FDDI and ISDN. It also defines the MAC sublayer and Physical layer interfaces. The 802.9 standard allows for internetworking between different subnetworks—networks that are a part of a larger network and are connected with bridges, routers, and gateways.

802.10: LAN/MAN Security

The 802.10 standard defines the assigning of unique Security Association Identifiers (SAIDs) for the purpose of security within and between LANs and MANs.

802.11: Wireless LANs

The 802.11 standard identifies a group of standards developed for wireless LAN technology. It defines communication between two wireless clients or a wireless client and an **access point**. It uses the 2.4GHz frequency band to transmit at up to 2Mbps. The 802.11 standard uses either frequency-hopping spread spectrum or direct-sequence spread spectrum. (Frequency-hopping and direct-sequence spread spectrum are explained in Chapter 8.)

802.11a: High-Speed Wireless LANs

The 802.11a standard goes beyond 802.11. 802.11a defines transmissions that utilize the 5GHz frequency band. It uses **orthogonal frequency division multiplexing** to provide up to 54Mbps of wireless throughput. (Orthogonal frequency division multiplexing is explained in Chapter 9.)

---

multiplexing
A process that allows multiple signals to transmit simultaneously across a single physical channel.

access point
A device that connects wireless users to a network.

802.11b: Most-Common Wireless LAN Standard

The 802.11b standard was adopted as the third 802.11 standard, and is the basis of most of today's wireless devices. Like 802.11, 802.11b uses the 2.4GHz frequency band for communication, although it allows for up to 11Mbps of throughput to be transmitted. This increased transmission rate made wireless networks more feasible in the average LAN environment.

802.11g: In the Works

In July of 2003, the IEEE is expected to adopt the 802.11g standard. It is currently in draft form, and is working its way through the approval process. To follow the progress of this work, view the reports and minutes of the meetings at `http://grouper.ieee.org/groups/802/11/Reports/`. When adopted, 802.11g is expected to transmit at up to 54Mbps.

802.12: High-Speed LANs

The 802.12 standards define how the Physical layer and MAC sublayer support 100Mbps signal transmission using the Demand Priority Access Method following the OSI reference model.

The Demand Priority Access Method puts the responsibility for transmissions on the hub. Devices request permission to transmit, and the hub determines the order of the transmissions and provides access to the network. Demand priority also allows for devices to be assigned a priority status so that their transmissions take precedence over other transmissions. This method allows for higher bandwidth transmissions between devices.

802.14: Cable TV Access Method

The 802.14 standard provides a reference for digital communications services over cable television networks using a branching bus system. The MAC and physical characteristics follow the 802 standards, including connectionless and connection-oriented communications.

802.15: Wireless Personal Area Network

This standard is also under development by the IEEE. It is focused on short-distance wireless networks called **WPANs** (**Wireless Personal Area Networks**).

WPAN
Wireless Personal Area Network. WPAN devices include PCs, PDAs (Personal Digital Assistants), cell phones, pagers, and much more.

Standards

These can include mobile devices such as PCs, **PDA**s (**Personal Digital Assistants**), cell phones, pages, and much more. The development of this standard will allow greater operability between these devices.

NOTE

Although it is much talked about, **Bluetooth** is not an official IEEE standard. It is the foundation for the 802.15 standard that is under development. It was developed by a group of manufacturers so that their devices could interoperate. Because it operates in the same frequency band as 802.11b, there can often be interference between devices transmitting simultaneously.

PDA
Personal Digital Assistant. A handheld device that can provide the functions of a personal organizer, phone, and fax sender, as well as beb browsing and other network features.

Bluetooth
A wireless technology developed jointly by companies such as Ericsson, IBM, Intel, and Nokia to support the exchange of data between wireless devices.

EIA/TIA Structured Cabling Standards

The EIA/TIA structured cabling standards are the guidelines used to design, install, and test network media and electrical infrastructure. The four most important of these are the 568, 569, 606, and 607 standards. The 568 standard has been revised and is now called 568-A.

568-A: Commercial Building Telecommunications Wiring Standard

The 568-A standard describes a basic system for the installation of cabling that can support a variety of purposes, including LANs and other telecommunications applications like telephone systems. This standard defines both hardware and wiring requirements.

TIP

The 568 standard was adopted in 1991. In 1995, it was updated and replaced by 568-A.

568-A/UTP

The 568-A/UTP standard specifies the application of unshielded twisted-pair (UTP) cabling, defining seven categories for UTP usage. Each category of cabling has its own requirements, including the wire's gauge, the amount of acceptable impedance, transmission speeds, and where it should be used. This standard also defines the colored wiring patterns that should be used for terminations, including plugs, jacks, punch blocks, and patch panels.

TIP

The 568-A and 568-B color schemes define the color order for twisted-pair wires.

Standards

569: Commercial Building Standard for Telecommunications Pathways and Spaces

In order for commercial buildings to provide the correct facilities to support telecommunications equipment and media, the 569 standard was adopted. The 569 standard sets requirements for data or wiring closets and pathways, both between and within buildings. The standard deals only with the telecommunication aspects of a commercial building.

Among other things, the 569 standard regulates:

- The appropriate size, location, and number of wiring closets
- Color codes for wiring components
- The separation distance between electrical and data wiring

This standard was adopted in 1990, before the cabling standards were adopted.

606: Administration Standard for the Telecommunications Infrastructure of Commercial Buildings

The 606 standard covers the marking and documentation of the wiring, pathways, and grounding found in a commercial building. Detailed diagrams are expected to include telecommunication outlets, closets, equipment, horizontal pathways, backbone pathways, and grounding.

In addition, the 606 standard requires that cable terminations (jacks and punch blocks) and wiring outlets are clearly identified and labeled. Larger and more complex buildings need much more detailed documentation than smaller buildings.

Correctly labeled Category 5 cable should have:

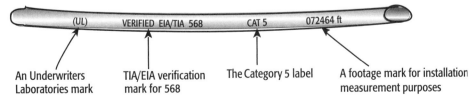

- An Underwriters Laboratories mark
- TIA/EIA verification mark for 568
- The Category 5 label
- A footage mark for installation measurement purposes

607: Commercial Building Grounding and Bonding Requirements for Telecommunications

The requirements for grounding and bonding are outlined in the 607 standard. Every commercial building must be properly grounded so that any excess voltage can dissipate. Without proper grounding, both devices and data are in jeopardy of being destroyed or damaged.

The 607 standard requires the following:

- There must be a reference ground in every building, including in the telecommunications closet and equipment room.
- The connection of cabling in the wiring closet meets specific guidelines for cable shields, conductors, and hardware.

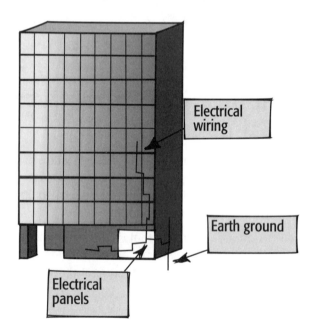

Standards

WAN Connection Standards

As you learned in Chapter 1, wide area networks interconnect local area networks. WANs allow users on LANs hundreds of miles apart to communicate and share important information and resources. The road to creating reliable WAN connections has not been smooth. Early on, the development of standard WAN technologies was hindered by independent companies that developed their own **proprietary** solutions.

> **NOTE**
>
> Proprietary equipment or software is developed by one company to support and be used by only their own equipment and technologies. It is owned by that company and is not available for use by the public or other companies without paying for the privilege. An example is Cisco's routing protocol EIGRP, which is exclusive to Cisco routers.

It isn't uncommon that companies create competing technologies that wind up being incompatible with one another. As this problem was true of early LANs, it was also true of early WANs.

In the 1950s and 60s, companies such as AT&T and its Bell Laboratories research group were responsible for installing the cabling infrastructure that would eventually make the first WANs possible. There was a need for a reliable and efficient method for connecting and disconnecting voice calls. The devices that were developed and installed were called *switches*, and they took the place of the early telephone operators who connected calls by hand. This switch-based system emerged as the **public switched telephone network (PSTN)**.

By introducing switches, the telephone company was able to accomplish two major goals:

- Improve the speed of connections using equipment instead of people
- Convert all connections between **central offices (COs)** to digital signals

The PSTN made it possible to connect calls across great distances without involving people. This new level of efficiency made **public data networks (PDNs)** possible.

CCITT/ITU-T WAN Standards

As we discussed earlier, the CCITT/ITU-T is responsible for a number of connection standards for wide area networks.

---

proprietary
The sole property of one individual or organization.

public switched telephone network (PSTN)
The term for the worldwide telephone networks and their services. Also known as *plain old telephone service (POTS)*.

central office (CO)
The local telephone company office in which all local loops connect and provide switching of telephone circuits.

public data network (PDN)
A fee-based network whose main concern is to provide computer communications to the public. It may be operated by a private company or a government.

X.25

In the 1970s, the international organization called the Consultative Committee on International Telephony and Telegraphy (CCITT) approved the **X.25** protocol for international use. It is considered to be the first WAN protocol.

The X.25 protocol has the following characteristics:

- It uses any available path to create a connection.
- It uses error checking and retransmission.
- The maximum bandwidth for transmission is 64Kbps.
- It has been widely adopted and is available worldwide.

X.25 is a packet-switching protocol that utilizes the public phone switches to establish a connection between two computers. X.25 does not require **leased lines**. Leased lines can be very expensive because you pay for the use of the line full time, whether you are always using it or not.

X.25
A wide area networking standard that defines how devices communicate within public data networks using data terminal equipment and data communications equipment.

leased line
A type of dedicated transmission connection for the use of a private customer.

virtual circuit
A logical circuit used in Frame Relay and X.25 WANs that provides communication between two network devices, and can be either permanent or switched.

Frame Relay
A WAN connection protocol that provides multiple virtual circuits. It is more efficient and less expensive than many WAN connections, and is currently being considered as the replacement for the X.25 protocol.

NOTE
You will learn more about leased lines in Chapter 14.

X.25 takes advantage of the phone switches by creating a **virtual circuit**, a connection that is maintained between two devices or customers (although communications between the devices may take an indirect path). For example, a virtual circuit is created when a call is made from San Francisco to Los Angeles. The initial call comes into the switch in San Francisco. The switch then establishes a connection through other switches until it reaches the location in Los Angeles. When the call is over, the connection to Los Angeles and all points in-between is terminated.

Frame Relay

The **Frame Relay** protocol was originally developed by CCITT in 1984. ANSI also conducted its own development through its T1S1 standards committee. People were excited about Frame Relay because of its capability to support speeds that exceeded that of X.25.

In 1990, Frame Relay received a major boost in its development due to a joint venture between Northern Telecom, Cisco Systems, Stratacom, and Digital Equipment Corporation. The Frame Relay protocol developed by this commercial partnership maintained the same specifications found in the CCITT standard, but it added additional features that enhanced the protocol.

Standards

Frame Relay has the following characteristics:

- It uses packet switching to transmit data.
- Data can be transmitted at speeds up to 45Mbps.
- There's no error correction.

Like X.25, Frame Relay is a packet-switching protocol. Frame Relay uses the same switching technology as X.25, but is capable of much greater speeds. Frame Relay can take advantage of digital lines on copper and fiber-optic cable at speeds up to 45Mbps. This far outperforms the 64Kbps limitation of X.25, so Frame Relay is well-suited for today's bandwidth-intensive applications.

Frame Relay can achieve much higher throughput due in part to lower overhead compared to X.25. Unlike X.25, Frame Relay doesn't utilize error correction or retransmission of garbled data. Frame Relay relies on the upper-layer protocols to provide these functions, a solution that was not available when X.25 was developed.

Integrated Services Digital Network (ISDN)
A WAN communications protocol offered by telephone companies that allows telephone lines to carry data, voice, and other types of transmissions.

> **NOTE**
> Frame Relay is a layer 2 protocol, and it encapsulates a layer 3 address, such as IP, within the Frame Relay packet.

Integrated Services Digital Network

Integrated Services Digital Network (ISDN) is another protocol adopted by ITU-T that uses digital signaling over regular phone lines. Because ISDN transmits digitally throughout the connection from computer to computer, it can provide a very high rate of transmission with excellent quality. Also, ISDN is capable of carrying voice and data for special purposes such as video teleconferencing.

ISDN has the following characteristics:

- It uses regular telephone lines.
- It can be used for voice and data.
- There are two types of ISDN:
    - ISDN BRI (Basic Rate Interface)—uses two B channels working at 64Kbps.
    - ISDN PRI (Primary Rate Interface)—uses 23 B channels of 64Kbps each.
- ISDN includes an additional single channel, the D channel, working at 16Kbps or 64Kbps.

Chapter 10

ISDN can support multiple services such as data, voice, and fax simultaneously due to the use of multiple **B channels**. ISDN BRI has two B channels that are each capable of transmitting at 64Kbps, for a combined total throughput of 128Kbps. In addition to the two B channels, there is a 16Kbps **D channel** that manages signaling of the B channels. The B channels can be combined to support only data, one data and one voice channel, or all voice. ISDN PRI, on the other hand, uses a 64Kbps D channel to manage the 23 B channels.

Unfortunately, ISDN has not had a very impressive track record. Because ISDN has not seen the popularity that was expected in the United States, the cost of ISDN has remained relatively high in comparison with other WAN technologies that provide similar transmission speeds. When the telecommunications companies implemented the original CCITT I.120 standard developed in 1984, they weren't coordinated in their efforts. This created difficulties when installing a reliable ISDN connection—a problem that persists to this day.

B channel
Bearer channel. A 64Kbps channel used in ISDN that provides full-duplex communication for the user.

D channel
Data channel. A 16Kbps (BRI) or 64Kbps (PRI) full-duplex ISDN channel.

Fiber Distributed Data Interface (FDDI)
A type of network that uses 100Mbps dual-ring token passing with fiber-optic cable.

> **TIP**
> WAN technologies will be discussed in detail in Chapter 14.

ANSI WAN Standards

The American National Standards Institute has adopted several WAN standards that have not yet been adopted internationally. They include FDDI, ATM, and SONET.

Fiber Distributed Data Interface

Fiber Distributed Data Interface (FDDI, pronounced "fid-ee") is defined as a token-passing network (similar to IEEE 802.5 Token Ring) over dual, redundant, counter-rotating rings using fiber-optic cable. FDDI is capable of transmitting data at speeds greater than 100Mbps. By using repeaters, FDDI connections can span over 100 kilometers.

FDDI has the following characteristics:

- It uses token passing.
- It uses a dual ring for fault tolerance.
- Data can be transmitted at speeds beyond 100Mbps.
- Networks can stretch over 100 kilometers or more.

Standards

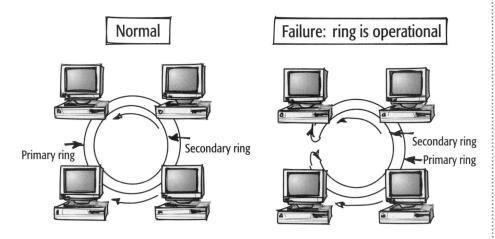

In the previous illustration, the primary ring represents the path that data travels under normal conditions. The secondary ring provides a level of fault tolerance and backup in cases where a connection may fail. In the event that a connection does fail, the secondary ring will provide a path to complete the ring, as shown on the right. This is known as *counter-rotating*, an important feature of FDDI. FDDI can provide excellent service for transmitting data, but it is very expensive. FDDI is being surpassed by technologies such as ATM that deliver voice, data, and multimedia information at faster speeds.

Asynchronous Transfer Mode (ATM)

Like Frame Relay, **Asynchronous Transfer Mode (ATM)** was developed because of demands for faster connections. Currently, ATM is a standard set by ANSI's T1S1 committee, and it is recognized only in the United States. Other organizations in Europe are researching and developing their own flavors of ATM. The ITU-T is making a major effort to establish a global standard for a technology similar to ATM, what they call **Broadband ISDN (BISDN)**.

ATM has the following characteristics:

- It's currently accepted only in the United States.
- Data can be transmitted at speeds from 25Mbps to 9,954Mbps.
- It uses fiber-optic cable.

ATM is a significant improvement over previous WAN technologies because of its incredibly high capacity. ATM is best suited to fiber-optic cable. ATM can transmit at speeds beginning at 25Mbps and can run as high as 9,952Mbps (9.952Gbps). An ATM network running at 9.9Gbps is capable of handling 155,500 simultaneous voice conversations—all over one cable!

Asynchronous Transfer Mode (ATM)
The transmission of digital signals using different frequencies and timing, allowing for multiple transmissions to occur through the use of packet switching.

Broadband ISDN (BISDN)
An extension of ISDN that is capable of supporting voice, data, and video at very high speeds using ATM as the transmission system.

Chapter 10

Although ATM has proven to be able to sustain a very high capacity, it does so at a price. ATM requires that telecommunications companies install fiber-optic cable and costly ATM switches. Fortunately, the multiservice capabilities of ATM to provide voice, data, audio, and video have excited telecommunications companies enough that they have been upgrading their infrastructure for several years now. Most of the U.S. telecommunications backbone (both voice and data) runs on ATM.

Synchronous Optical Network (SONET)
A high-speed synchronous network that runs on optical fiber at speeds of up to 9.9Gbps.

Synchronous Optical Network

Synchronous Optical Network (or SONET, pronounced "saw-net") is a high-speed data-transmission system. SONET is a Physical layer implementation for transmitting information across optical networks at speeds beginning at 51.84Mbps and reaching as high as 9.9Gbps. For this reason, SONET is often the underlying technology that allows ATM networks to exist. The various SONET line speeds are classified as **OC (optical carrier) levels**. (For technologies that support protocols such as X.25 and Frame Relay, the different rates are known as *digital service levels*.)

OC (optical carrier) levels
Line speeds in SONET networks.

NOTE

SONET networks are not commonly implemented today. As the technology becomes less cost-prohibitive, they will likely become more popular.

Internet 2
The second Internet, being implemented between academic research organizations using high-speed connections.

SONET has the following characteristics:

- It transmits across optical networks.
- Transmission speeds range from 51.84Mbps to 9.9Gbps.
- It uses OC levels to define the speed and the number of channels.

NOTE

When companies need many direct phone lines, they order a DS-1 line from the telephone company that provides 24 phone lines on one cable.

The optical carrier levels used in SONET are shown in the table.

The University of California is using SONET in an OC-192 network connection between San Francisco and Los Angeles as part of the **Internet 2** project.

Standards

SONET Level	Transmission Rate
OC-1	51.84Mbps
OC-3	155.52Mbps
OC-9	466.56Mbps
OC-12	622.08Mbps
OC-18	933.12Mbps
OC-24	1.244Gbps
OC-36	1.866Gbps
OC-48	2.488Gbps

NOTE

Internet 2 is currently being developed to provide high-speed intercommunication between academic research institutions. It is fully detailed at www.internet2.edu.

Review Questions

Terms to Know
- American National Standards Institute (ANSI)
- Asynchronous Transfer Mode (ATM)
- B channel
- backoff
- Broadband ISDN (BISDN)
- Carrier Sense Multiple Access/Collision Detection (CSMA/CD)
- cell
- central office (CO)
- D channel
- Electronics Industry Alliance (EIA)
- Fast Ethernet
- Fiber Distributed Data Interface (FDDI)
- Frame Relay
- Institute of Electrical and Electronics Engineers (IEEE)
- Integrated Services Digital Network (ISDN)
- International Organization for Standardization (ISO)
- International Telecommunications Union (ITU)
- Internet 2
- leased line

1. Why are standards important in networking?

2. What are the six primary organizations that adopt networking standards?

3. What does ISO stand for, and what is the primary standard that it adopted related to networking?

4. What is the IEEE 802.3 standard, and how does it work?

5. What is a backoff?

6. What is 803.2u?

7. What is 802.3z?

8. What access method does Gigabit Ethernet use?

9. What do the 802.11 standards support?

10. Which is the most common wireless LAN standard today?

11. What wireless standard supports connectivity between most mobile devices?

_____

12. What is the 802.5 standard, and how does it work?

_____

13. What is the EIA/TIA?

_____

14. What does the EIA/TIA 569 standard define?

_____

15. Why is the 607 standard important to data transmission and devices?

_____

16. What does PSTN stand for?

_____

17. What is the only internationally adopted WAN protocol?

_____

18. What does ATM stand for?

_____

19. How does FDDI work?

_____

Terms to Know
- ❏ Multiple Access
- ❏ multiplexing
- ❏ OC (optical carrier) levels
- ❏ orthogonal frequency division multiplexing
- ❏ Personal Digital Assistant (PDA)
- ❏ proprietary
- ❏ public data network (PDN)
- ❏ public switched telephone network (PSTN)
- ❏ Request for Comment (RFC)
- ❏ Spanning Tree Protocol (STP)
- ❏ Standard
- ❏ Subnetwork Access Protocol (SNAP)
- ❏ Synchronous Optical Network (SONET)
- ❏ Telecommunications Industry Association (TIA)
- ❏ token
- ❏ virtual circuit
- ❏ Wireless Personal Area Network (WPAN)
- ❏ X.25

Chapter 11

Network Protocols

Protocols are similar to standards. They set the rules and conventions that define how devices on a network exchange information. Within network communications, protocols define how two devices will carry on their conversation.

A single protocol can't provide complete intercommunication independently. It must work with other protocols, operating at different layers of the OSI model, to provide complete end-to-end communication. When a set of protocols works together, it is called a *protocol suite* or *stack*. There are three popular protocol suites that are common in today's local and wide area networks: TCP/IP, IPX/SPX, and AppleTalk.

In this chapter, you'll learn:

    The importance of protocols

    Why CCNAs need to understand protocols

    The foundations and features of the three main protocol suites

    Individual protocols that function within the popular protocol suites

Chapter 11

Why Protocols Are Important

As you learned in Chapter 10, standards define how devices communicate and access media. Technology companies follow these standards when they develop devices or applications for network communication. Once a standard is implemented in software, it becomes a protocol.

Protocols define how devices and even applications communicate. Prior to the development of protocols, there was no intercommunication between devices. The protocols used on your network impact both how the network functions and its ability to communicate with other networks. (Your "choice" of protocols may be based on the type of network operating system you implement and may not be something you can change.)

One protocol that is available on all Windows machines is NetBEUI. It is one of the fastest communication protocols used today, but it is not routable. It cannot be passed by a router and can only be used within an internetwork. A common protocol suite, IPX/SPX, is a part of Novell's system of communication within a NetWare client-server environment. It has many benefits and features and can be routed, but only to communicate with other NetWare networks and not on the Internet.

NOTE

Protocol suites are also commonly referred to as *stacks* because they include individual protocols that work together at different layers of the OSI model.

As a Cisco Certified Networking Associate, you will encounter many types of protocols being implemented within a variety of network environments. You may see legacy IBM mainframes using SNA to communicate, or a computer lab in a school running AppleTalk. When working with these or any other systems, your job will most likely be to ensure that these networks can communicate with each other as well as with other often very different systems. To do this, you must understand how the protocols being used work and how they impact the network's performance.

You will also be responsible for implementing routing protocols to pass the routed protocols being used on your network. (You learned about routing and routed protocols in Chapter 9.) In some instances, you may need specific hardware or an internetwork operating system (IOS) that can pass specific protocols between routers. Not all routers can pass all types of protocols.

Network Protocols

In addition to knowing that you can program your router to pass the type of protocol you are using, you must also be aware of the impact different routing protocols have on the way data is passed between routers.

Finally, as you design and install an internetwork, understanding protocols also allows you to implement added security and reduce network traffic by managing specific protocols via router interfaces. This in turn allows for increased functionality. This level of sophistication requires an understanding of both individual protocols and protocol suites.

As you continue your training after the CCNA and work toward a Cisco Certified Networking Professional certificate, you will need to understand more complex ways of managing protocols and their impact on the network's performance.

NOTE

To receive CCNP certification, you must pass four exams. You'll find specific certification information in the introduction to this book.

Understanding Protocol Suites

The three most important protocol suites used in internetworking are:

- Transmission Control Protocol/Internet Protocol (TCP/IP)
- Internet Packet eXchange/Sequence Packet eXchange (IPX/SPX)
- AppleTalk

Each of these protocol suites implements a set of rules that provides a unique method for intercommunication between devices. TCP/IP is the most common of all network protocol suites and is the standard in today's networks. It is the protocol suite used for communication on the Internet.

IPX/SPX, developed by Novell, uses its own methods to ensure communication between devices using the NetWare operating system. Each of the individual protocols within the stack may provide a similar function to those in the TCP/IP stack, but they are proprietary to Novell.

AppleTalk, although often neglected, is still implemented and used in many networks, especially within the educational arena. It is the protocol suite used for communication between devices using the Mac OS.

The protocols within each suite function at different layers of the OSI model from layer 2 (the Data Link layer) to layer 7 (the Application layer). This chapter will identify and explain the protocols found at the different layers of the OSI model for each of the three protocol suites:

- Data Link layer protocols
- Network layer protocols
- Transport layer protocols
- Application layer protocols

NOTE

Since the TCP/IP and IPX/SPX protocol suites were developed before the adoption of the OSI model, they do not map perfectly to its layers.

Network Protocols

TCP/IP Suite

The **TCP/IP suite** was developed for use on the Internet. Its origins lie in the first RFC, which was discussed in Chapter 3. It is also known as the DoD or ARPAnet protocol suite. Its name comes from two of the main protocols within the stack: Transmission Control Protocol and Internet Protocol. TCP is responsible for connection-oriented communications using error checking, and IP is implemented in the addressing system used to identify devices.

TCP/IP suite
The protocol suite developed to provide access to the Internet.

Although developed for use on the Internet, TCP/IP is implemented to build LANs, MANs, and WANs. TCP/IP is the most widely implemented protocol suite and is used within a variety of platforms, including UNIX, Windows, and the Macintosh. The following illustration maps the protocols of the TCP/IP protocol suite to the layers of the OSI model:

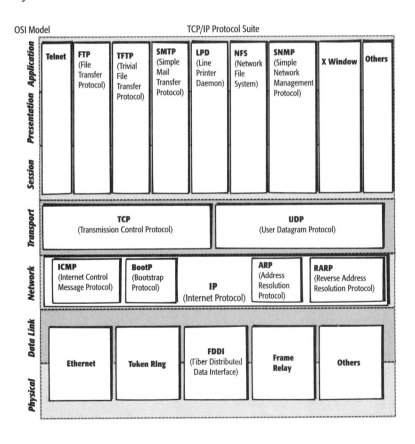

237

Features of TCP/IP

The TCP/IP suite offers a number of features and benefits, which are outlined in the next sections.

Interoperability

TCP/IP has become the standard protocol suite because of its capability to connect LANs, WANs, and the Internet. As more LANs connect to the Internet, TCP/IP is being installed more frequently. And in the latest version of its NetWare operating system, Novell has replaced two of its proprietary protocols, IPX and SPX (which are discussed later in this chapter), with TCP and IP in the NetWare protocol suite. TCP/IP is quickly becoming the most universally available protocol today.

Flexibility

The multiple protocols within the TCP/IP suite provide for a variety of implementations. The choice between TCP, a connection-oriented method of communication that is reliable but slow, and UDP, a fast and efficient yet not as reliable method, is important when determining how a packet should be sent.

Multivendor Support

Another big benefit of the TCP/IP suite is the fact that almost all network software vendors support its use. Apple, DEC, IBM, Novell, Microsoft, and Sun are just a few of the many companies that support the suite.

Protocols of the TCP/IP Stack

The easiest way to break down the many protocols of the TCP/IP suite is according to where they operate within the OSI model. Each protocol in this stack operates at one of four layers: the Data Link layer, the Network layer (also known as the Internet layer in the TCP/IP model), the Transport layer, or the Application layer.

Data Link Layer Protocols

The protocols that operate at the Data Link layer define the access method for the media and architecture and interface with the Physical layer of the network. These protocols can be divided into LAN protocols and WAN protocols. These protocols are based on the standards that have been adopted by the IEEE as the 802 standards.

Network Protocols

> **NOTE**
> Standards, such as the IEEE 802 standards, are covered in Chapter 10.

LAN protocols that have been introduced in previous chapters include Ethernet, Token Ring, and FDDI. In Chapter 14, you will learn about WAN protocols such as PPP, ISDN, and Frame Relay.

Internet (or Network) Layer Protocols

In the TCP/IP protocol suite, the Network layer is known as the Internet layer. The Internet layer is responsible for the logical addressing of devices on the network, as well as routing data between a source and destination. The four most important TCP/IP protocols that function within this layer are:

- Internet Protocol (IP)
- Address Resolution Protocol (ARP)
- Reverse Address Resolution Protocol (RARP)
- Internet Control Message Protocol (ICMP)

Internet Protocol   Internet Protocol (IP) is one of the most important protocols you'll learn about. It was originally developed to function in the UNIX environment in the days of ARPAnet when UNIX was the only operating system. IP uses connectionless delivery, which means it does not guarantee delivery to the destination. Its main purpose is to provide logical addressing through the use of an IP address. This address is the identification number used to route information between networks using TCP/IP and within the Internet, and it is unique for every device on a network.

> **NOTE**
> The IP protocol is probably the most important protocol to understand. It defines the addressing system that connects all devices to the Internet. Without it, we would not be able to access anything on the World Wide Web.

Address Resolution Protocol   While an IP address is necessary to route data between networks in the TCP/IP environment, the physical address, called the *MAC address* in an Ethernet network, is also necessary to provide intercommunication between devices. It is not always possible for a source device to know the physical address of the destination device.

Address Resolution Protocol (ARP) provides the service of matching a known IP address for a destination device to a MAC address. ARP sends a **MAC broadcast** request to the entire destination network, asking for the MAC address of a particular known destination IP. The device that is identified by the IP address on the packet responds to the request, and replies by sending its MAC address. Thus, the source device is capable of correctly addressing communication packets without having to broadcast all messages and slow down the network.

MAC broadcast
A broadcast packet sent on the network with the MAC address of FF-FF-FF-FF-FF-FF, which requires all devices to pass the packet to the Network layer for address identification.

Reverse Address Resolution Protocol   The Reverse Address Resolution Protocol (RARP) provides the reverse service to that of ARP. Rather than finding out the MAC address of a device whose IP address is already known, RARP provides the IP address for a device that knows its own MAC address.

This service is typically needed in networks that use diskless workstations. These devices cannot maintain their own IP address in memory, but do have access to their MAC address. As a result, they must request an IP address to communicate.

A device known as a RARP server responds to the RARP request, and provides an IP address for the workstation or device to use while it is utilizing the network. Once the device is turned off, the IP address is available for the RARP server to provide to another device.

Internet Control Message Protocol   Internet Control Message Protocol (ICMP) is a valuable protocol for CCNAs and other network administrators. It is implemented in all TCP/IP networks, and provides messaging that can help with troubleshooting. ICMP messages are included in the IP datagram. The most common error and control messages sent include:

- Destination unreachable
- Time exceeded
- Redirect
- Echo
- Echo reply
- Information request
- Information reply
- Address mask request

Network Protocols

Understanding IP Addressing

A device's unique IP address that is encapsulated onto a data packet is made up of 4 bytes, or 32 bits. A sample device IP address using four-byte dotted notation is 209.100.29.83.

An IP address consists of two parts: the network portion (or number) and the host portion (or number). The network portion is used to route data packets between networks, whereas the host portion determines the destination device within the network.

The network portion is the part that is given to you by an Internet service provider. In the past, this number was requested directly from InterNIC. There are five classes of networks that can be identified based on the first bits of their address. The following illustration shows the class of network, the bits in the first octet that define it, and the number of bits that are included in the network and host portions of the address. The last two classes are reserved for special purposes: Multicast addresses, Class D, are used for groups of hosts that are confined to a limited area. Class E is reserved for experimentation and future use.

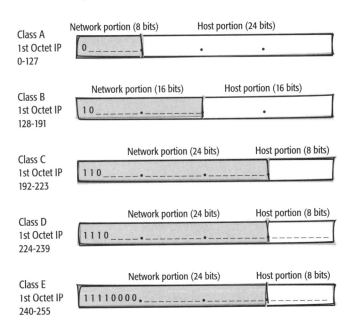

Although network numbers are assigned by the ISP, it is the job of a network administrator to assign the IP addresses to individual devices. This can be done manually or dynamically using Dynamic Host Configuration Protocol, which is explained later in this chapter. It is only the host portion of the address that can be changed and assigned; the network number is consistent throughout the network.

The total number of IP addresses that are possible by using this system is 4,294,467,295. This may seem like a lot of addresses, but we are running out. This current version of IP is version 4. RFC 1166 describes IP addresses and addressing. You can view the proposals at http://rfc.sunsite.dk/.

IP Subnets

Networks, especially Class A and Class B networks, must often be divided into smaller networks, called **subnetworks** (or subnets). To do this, a network administrator must use the host portion of an address to create smaller groups of addresses. This is done through a process called **subnetting**.

A **subnet mask** is created when the administrator "borrows" bits that are a part of the host address portion and uses them to create subnets. The number of bits that are borrowed defines the number of subnets that are generated. The number of bits that are left in the host portion identify the number of host IP addresses in each subnet.

The parts of a subnetted network number are divided like this:

<network portion><subnetwork portion><host portion>

The following illustration shows an example of a static subnet being used with a Class C network address.

Internet Protocol Version 6

The new version of the Internet Protocol that is being developed—Internet Protocol version 6—is currently called IPv6. For addressing, it will use eight groups of hexadecimal numbers of 16 bits each, for a total of 128 bits. It uses hexadecimal numbers to represent the 128-bit address as well as colons between bytes, so it looks much different from the format we are used to for IP addressing. Hopefully, this version of the Internet Protocol will provide sufficient addresses for our future needs.

An IPv6 address will look like this:

3524:72D3:2280:A302:0029:37C4:F2E2:002B

---

subnetwork (subnet)
A smaller segment of a larger network that is created by a network administrator through the use of a subnet mask.

subnetting
A method of splitting an IP network address into smaller groups of IP addresses that can be used on different networks.

subnet mask
A special 32-bit address used to indicate the bits of an IP address that are being used to create subnets. Sometimes called a *subnet address* when shown as a decimal dotted notation number.

Network Protocols

Transport Layer Protocols

Transport layer protocols have the job of providing end-to-end communication between devices. The key function of the protocols at this layer is reliable and efficient delivery of data packets to the destination. In TCP/IP, there are two Transport layer protocols that provide the end-to-end communication services:

- Transmission Control Protocol (TCP)
- User Datagram Protocol (UDP)

One of the most important distinctions between them is that TCP uses connection-oriented services, whereas UDP uses connectionless delivery.

Transmission Control Protocol The connection-oriented services of the Transmission Control Protocol (TCP) ensure the delivery of packets. TCP uses **checksums** that are added to the data packet so that the destination device can determine if the packet was received without any errors. By matching the information in the checksum to the data that is received, that destination knows whether the packet is intact. TCP's error checking is also aided by the use of sequence numbering, which identifies the order in which the packets were sent.

The destination device is required to acknowledge the receipt of each packet sent. If the device was received, it asks that the next packet be sent. If the packet was damaged, the destination requests that it be retransmitted. This connection-oriented system also creates additional traffic on the network, and can have a negative impact on network performance.

User Datagram Protocol The User Datagram Protocol (UDP) uses a connectionless transportation system. It does not use either error checking or sequence numbering to help in the guarantee of delivery. UDP assumes that other protocols and layers within the stack will provide the error checking it does not. It is concerned only with getting the data to the Transport layer of the destination device. It does not know or care what the other layers do with the data. Because UDP does not require that the destination device acknowledge each packet that is sent, this protocol requires fewer network resources.

UDP uses less bandwidth than TCP, which makes it more efficient but less reliable as an independent protocol. But combined with other protocols in the suite, it provides the necessary services for communication to occur.

Application Layer Protocols

Application layer protocols fall between the Session layer and Application layer of the OSI model. (As you learned in Chapter 2, because the TCP/IP and IPX/SPX protocol suites were developed before the OSI model, their layers do not perfectly match those of OSI.) The Application layer protocols provide network services to

> **checksum**
> A way of providing error checking by calculating a value for a data packet that can be used by the destination device, which recalculates the value of the packet and can determine if the packet was received intact.

the end user, such as e-mail and file transfer. The TCP/IP protocol suite includes many Application layer protocols. You should know what these protocols do, because you will find them in common networking applications and operating systems.

File Transfer Protocol File Transfer Protocol (FTP) is not only a protocol; it is also a service and an application. FTP is an important protocol that allows files to be copied between devices. Files may be downloaded (taken from another device) or uploaded (placed on a distant device). FTP was one of the first protocols developed; it is outlined in RFC 454.

Trivial File Transfer Protocol Like FTP, Trivial File Transfer Protocol (TFTP) is used for the transferring of files, although specifically over the Internet. It works with UDP and provides its own errorchecking and acknowledgment system. TFTP requires an acknowledgment for each packet that has been sent before the next will be transmitted. This way, the connectionless transmissions of UDP can be accounted for.

Simple Mail Transfer Protocol Simple Mail Transfer Protocol (SMTP) has the responsibility of transferring e-mail between computers. It uses the connection-oriented services of TCP to send and receive messages. Not all operating systems understand SMTP because it was designed specifically for use with UNIX. As a result, SMTP gateways are needed to provide translation services.

HyperText Transfer Protocol HyperText Transfer Protocol (HTTP) is used to access **HTML (HyperText Markup Language)** files on the Internet. It allows for clients and servers to exchange data very rapidly. HTTP offers only two types of messages to be exchanged: requests from the client and responses from the server. It does not provide any security or memory of transactions that have taken place between the client and server. It can get files or send files, and that's about it.

HTTP has had a huge impact on the Internet. Today, there are new versions, such as Secure HTTP (SHTTP), that offer the security that was lacking in the original protocol through encryption and security checks.

Domain Name System The Domain Name System (DNS) has been implemented as a way to manage and centralize domain names on the Internet. It provides the service of matching a host name for a unique domain such as www.sybex.com to the IP address of a server where the host is located.

A domain name is made up of at least three levels. The top-level domain is identified by the last characters of the domain name; for example, .com (for a commercial organization), .org (for a non-profit), or .jp (for Japan). Recently, several new top-level domains were added in the United States to meet ever-increasing demands. The second level in a domain name like sybex.com is the actual name of the organization. The third level—the www—typically refers to a

---

HTML (HyperText Markup Language)
A scripting language, developed in the early 1990s, which uses a browser for viewing documents on the Internet.

Network Protocols

host, which is usually a server. It is also possible to have additional levels that are added to further distinguish between locations or departments in an organization; for example, www.cs.sfsu.edu.

Telnet   Telnet is both a protocol and an application. Its name comes from the way most Telnet sessions take place—over a telephone network. The Telnet protocol is used to provide remote terminal services. This allows an end user using a computer that is acting as a terminal to interact with remote devices and have a communication session with a program on the remote server. The end user uses their display and keyboard to interact with the remote server as if they were at the same physical location as the server. Telnet allows many users to access the server from remote locations at the same time.

Telnet software is designed to translate commands on a remote device so that the remote server can understand them. Telnet applications implement the Telnet protocol, which can translate the device's communications and commands into a generic format that the server can understand.

Dynamic Host Configuration Protocol   Dynamic Host Configuration Protocol (DHCP) dynamically provides IP addresses and other configuration information to devices requesting to utilize the network. Like RARP, DHCP will provide an IP address to a device that doesn't have one. This service, usually provided by a router or dedicated server, has become increasingly important as the Internet outgrows the current IP addressing system. DHCP allows a pool of IP addresses, based on the assigned network address, to be leased to a device. Depending on how the protocol is implemented, the lease can last either a specified amount of time or for the duration of the device's use of the network.

Simple Network Management Protocol   Simple Network Management Protocol (SNMP) offers the ability to configure, monitor, and manage network resources and devices. As with the other application protocols, software is needed to use SNMP, and all the devices on the network must be able to understand the protocol. You'll learn more about SNMP in Chapter 13.

IPX/SPX Protocol Suite

The **IPX/SPX** protocol suite was developed for use with Novell NetWare networks. Its name comes from the two most important protocols of the suite: Internet Packet eXchange (IPX) and Sequence Packet eXchange (SPX). While IPX and SPX have been widely implemented in Novell NetWare networks, the latest version of NetWare offers TCP and IP and a full suite of TCP/IP applications, thus providing Internet capabilities lacking in IPX and SPX.

IPX/SPX networks communicate using a proprietary protocol suite based on a modification of the **Xerox Network System (XNS)**. Like TCP/IP, IPX/SPX predates the development of the OSI model and does not match the layers of the OSI. As you can see in the following illustration, it more closely resembles the TCP/IP model than the OSI model.

IPX/SPX
Novell's proprietary protocol suite that provides end-to-end communication on Novell NetWare networks.

Xerox Network System (XNS)
One of the first protocols used with Ethernet, XNS became the model used by many vendors for developing protocols.

Features of IPX/SPX

The following sections discuss some of the features and benefits of IPX/SPX.

Network Protocols

Ease of Addressing

The IPX/SPX logical addressing system is much more user-friendly than the Internet Protocol addressing system. IPX uses a network number that is chosen by the network administrator and the Data Link (MAC) address of the device. These two numbers make up the entire IPX address. Many people feel that the IPX addressing system is much easier to implement than the IP addressing system, in which addresses must be requested from an ISP.

Get Nearest Server (GNS)
A process that uses SAP broadcasts to request services from servers available on the network.

> **NOTE**
> Although the network number in an IPX address may be chosen by an administrator, when configuring a Cisco router, you must use the IPX network address of a connected IPX network, if one exists.

Dynamic Service Discovery

One of the key features of the IPX/SPX protocol suite is the use of service advertisements. Service Advertisement Protocol (SAP) allows servers to advertise the services they can provide with broadcasts. These advertisements occur every 60 seconds by default. This does add traffic to the network.

This system of advertisements allows for nodes to discover which server they need to communicate with for specific services. Through a process called **Get Nearest Server (GNS)**, clients send out broadcast messages on the network in an attempt to locate the nearest server. Servers respond to the request with a GNS response. This way, servers that are nearest to the client and that have the needed services can be used.

SAP broadcasts are not routable. A router that receives a SAP broadcast records it in a SAP table. This SAP table is the key to providing services outside of a local area network. Routers share these SAP tables with each other. A device making a service request may receive a reply from a router that knows how to find the requested services, even beyond the local network where the client is located.

Protocols of the IPX/SPX Stack

As with TCP/IP, an easy way to understand the relationship between the various protocols of the IPX/SPX suite is to look at what layers they work in.

Network Layer Protocol: Internet Packet eXchange

The Network layer protocol implemented in Novell NetWare versions 2 through 4.3 is IPX. Like IP, IPX is responsible for network addressing. It uses a 12-byte address: up to four bytes for the network address and six bytes for the device's MAC address. The last two bytes are used for the socket number. The IPX header includes all the information needed by the upper-layer protocols that reassemble the data packets in sequence. The latest version of NetWare, version 5, offers the option of running a purely TCP/IP server along with Internet server applications or IPX/SPX.

Transport Layer Protocol: Sequence Packet eXchange

Novell's proprietary Transport layer protocol, SPX, works with its IPX protocol. Like TCP, SPX uses connection-oriented delivery. To ensure reliable delivery, all packets are sent in order and contain sequencing information so that they can be reassembled. SPX is also responsible for the flow control between end systems. (Flow control is discussed in Chapter 2.)

Application Layer Protocols

The proprietary protocols for Novell NetWare networks provide important services between the server and devices such as workstations. These services include printing and file handling, as well as advertising the services available from a specific server.

Service Advertisement Protocol As explained earlier, through the use of Novell's Service Advertisement Protocol (SAP), servers on a NetWare network can use broadcasts to make their services known to the network.

NetWare Core Protocol Novell's NetWare Core Protocol (NCP) provides services that span the top three layers of the OSI model. Its function is to provide NetWare client-server connections. NCP gives many important capabilities to the server, which in turn provides services to the workstations. The services that are provided by NCP include establishing and ending a connection, file services, printing, and security. NCP can also bypass SPX, the layer 4 protocol, and can communicate directly with IPX for communication that is connectionless.

AppleTalk Protocol Suite

AppleTalk is the proprietary protocol suite developed in the mid-1980s for use with Apple Macintosh networks. It is a multilayer architecture that is built into the Macintosh operating system. Because of this, all Macintosh computers are capable of networking right out of the box. Recently, changes have been made to the suite to improve its communication capabilities.

Unlike the TCP/IP and IPX/SPX protocol suites, the AppleTalk protocol suite can be mapped to the OSI model.

> **AppleTalk**
> Apple's proprietary protocols that are built into the Mac OS and allow Apple Macintoshes to communicate right out of the box.

Features of AppleTalk

AppleTalk also has some features and benefits worth noting.

Ease of Addressing

The network portion of an AppleTalk address is manually configured by the network administrator, although device numbers are dynamically assigned, using eight bits or the numbers 1–253. The numbers 0, 254, and 255 are reserved.

This dynamic addressing happens when a device is turned on. A device chooses a random number from the range available and sends out a message asking if it is being used. If a device replies that it is using that address, another address is tried. If no one responds to the advertisement of this number, the device maintains it as its address. It then saves this address in RAM.

Limiting Traffic

Devices located within an AppleTalk network are assembled into logical zones. These zones block the broadcasts that are sent within the network. Each zone can then be connected to an interface on a router. The router can block the broadcasts, but can provide information about all of the connected zones to the user. In this way, traffic is limited while access is provided to all devices throughout the network.

NOTE

AppleTalk is the "chattiest" of all protocols. Devices using AppleTalk send broadcast messages every 10 seconds. These constant broadcasts, which can be limited only by layer 3 devices such as routers, dramatically impact network performance. If you implement an AppleTalk network, be sure to limit the number of devices within the broadcast domain by using a router.

Chapter 11

Protocols of the AppleTalk Stack

AppleTalk protocols were designed as a client-distributed network system to provide the sharing of resources within the Macintosh environment. As with SAP in the IPX suite, AppleTalk clients use broadcasts to find out what services are available to them. A logical group of devices within an AppleTalk environment is called a **zone**.

> **zone**
> A group of network devices within an AppleTalk network.

There are currently two versions of AppleTalk: Phase 1 and Phase 2. Phase 1 AppleTalk supports one physical network with one logical network or zone. Phase 2 supports one physical network with more than one logical network and multiple zones.

The next sections discuss the AppleTalk protocols specific to particular layers, with some described in detail.

Data Link Layer Protocols

AppleTalk has developed its own Data Link layer protocols, which are designed to provide services similar to the 802 standards for their proprietary system. As you can see from their names, their purpose is to provide the access control for the various types of network available. These protocols include:

- EtherTalk Link Access Protocol (ELAP)
- AppleTalk Address Resolution Protocol (AARP)
- LocalTalk Link Access Protocol (LLAP)
- TokenTalk Link Access Protocol (TLAP)

AppleTalk Address Resolution Protocol   AARP is utilized only when a physical address is needed to transmit to a destination.

LocalTalk   LocalTalk is Apple's Data Link and Physical layer protocol. It can handle multipoint connections utilizing a physical bus topology. Although it is slow, LocalTalk allows for a dynamic addressing scheme that allows workstations to join the network at startup. It is limited to 32 devices within the workgroup.

AppleTalk Middle-Layer Protocols

Following the OSI model, the AppleTalk network protocols that function at the OSI's middle layers include:

- Datagram Delivery Protocol (DDP)
- Routing Table Maintenance Protocol (RTMO)

Network Protocols

- Name Binding Protocol (NBP)
- AppleTalk Transaction Protocol (ATP)

Datagram Delivery Protocol

The Datagram Delivery Protocol provides Apple networks of all network architectures with a best-effort delivery system. It does not guarantee delivery of the data. Two types of DDP packets are used: Short DDP and Long DDP. Short DDP is used when packets are being sent within a network. Long DDP is used for communication between networks.

Name Binding Protocol

Like the Domain Name System used on the Internet, Name Binding Protocol (NBP) matches device names to network addresses. This service is transparent to the user and allows for the use of logical naming conventions for devices.

AppleTalk Transaction Protocol

The AppleTalk Transaction Protocol (ATP) provides the necessary transport reliability between devices. It provides these services during a transaction between a requesting application or node and the responding program or node.

AppleTalk Upper-Layer Protocols

Apple's upper-layer protocols work at the session layer and above. The upper-layer AppleTalk protocols include:

- AppleTalk Data Stream Protocol (ADSP)
- AppleTalk Session Protocol (ASP)
- Printer Access Protocol (PAP)
- Zone Information Protocol (ZIP)
- AppleTalk Filing Protocol (AFP)

AppleTalk Data Stream Protocol

ADSP performs at both the Transport and Session layers of the OSI model. It is responsible for establishing connections, sequencing, and flow control. ACSP is an alternative to ATP (mentioned previously) at the Transport layer, although ADSP does not keep track of transactions.

Zone Information Protocol

ZIP is responsible for keeping track of network numbers and zones. It matches the network number to the appropriate zone within an AppleTalk network.

AppleTalk Filing Protocol

As its name implies, AFP was developed to support file sharing. It provides end-user services, such as printing and file serving, both within and outside of the Macintosh environment. It also provides services between clients and servers.

Review Questions

1. What is a protocol?

2. What is a protocol suite?

3. How many layers are represented in the TCP/IP protocol suite, and what are they?

4. What is IP responsible for?

5. What is a subnet mask used for?

6. What is ARP, and how does it work?

Terms to Know
- AppleTalk
- checksum
- Get Nearest Server (GNS)
- HTML (HyperText Markup Language)

Review Questions

Terms to Know
- ❑ IPX/SPX
- ❑ MAC broadcast
- ❑ subnet mask
- ❑ subnetting

7. When is RARP used?

_____

_____

8. What service does TCP provide?

_____

_____

9. How is UDP different from TCP?

_____

_____

10. What is FTP used for?

_____

_____

11. What is SMTP responsible for?

_____

_____

12. What service does DNS provide?

_____

_____

13. How does DHCP provide addressing to network devices?

_____

_____

14. How is SNMP helpful to network administrators?

_____

_____

15. What impact does the AppleTalk protocol have on a network?

_____

_____

16. How is an IPX address created?

_____

_____

17. What is AppleTalk's Zone Information Protocol used for?

_____

_____

Terms to Know
- ❏ subnetwork
- ❏ TCP/IP suite
- ❏ Xerox Network System (XNS)
- ❏ zone

Chapter 12

LAN Design

The preceding chapters have been dedicated to understanding the technology that makes networks work. As you have learned, there are several factors that must be considered to ensure that the network is able to provide the services and resources an organization needs. Depending on the size of the organization, the network could be very complex. To successfully build and manage a complex network, it requires that the technology be correctly aligned with the needs and ultimately the business goals of the organization.

Planning a network is an important part in the success of almost every organization. During the planning stages of a network, it is necessary to determine the long-term needs of the organization. These may include planning for the relocation of people and equipment, changes in the way the network is used, and expansion of the network to accommodate more users. Errors in planning can cost an organization tens of thousands of dollars later on. When planning and designing networks, the future costs for changes should be taken into serious consideration.

In this chapter, you will learn the steps taken and tools used in LAN design, including:

   The planning process

   Needs assessment

   Architecture, topology, and device selection

   Implementation considerations

Preparing for LAN Design

LAN design requires attention to detail, especially in the area of documentation. Don't be surprised if you accumulate hundreds of pages of information. But before collecting all this information, it is critical that you get organized. To better understand LAN design, it helps to approach it as a process. This process is represented in the following illustration:

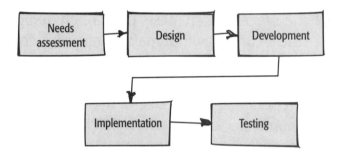

The process shown here looks much simpler than it really is. Each phase serves a distinct purpose and is dependent on the other phases.

Phase I—Needs Assessment During the needs assessment phase, user feedback, research on technologies and business practices, and other information are collected. The focus in LAN design is on how the existing network is being used, what the company's future plans are in terms of growth and computer usage, and what difficulty users have experienced in the past. All the information gathered here is used in the next phase.

Phase II—Design Once the needs of the company have been identified, the process of designing the network begins. The network engineer has the responsibility for creating a design that remedies network problems, allows for growth, and accommodates current and future network uses. In short, a successful design must be flexible.

Phase III—Development Although you have moved on to the next phase in the process, designing is still occurring. The goal in this phase is to minimize the possibility of a major change in the design during the implementation phase. The development process allows for changes to the design based on user feedback, testing of technologies included in the design, and any other new information that may require the original design to change. It is the job of the network engineer to take the feedback and problems discovered and adapt the design appropriately.

Phase IV—Implementation Now that the design has been developed and approved, it is time to install the network. On large projects, this may involve internal Information Technology staff, as well as external consulting companies.

LAN Design

In some cases, it may be the responsibility of the network designer to coordinate the network installation with IT consultants, the project manager, and any other staff that may be required. On smaller projects, the network designer may need to provide only a detailed diagram and equipment specifications to the cable installers.

Phase V—Evaluation   In this phase, the network is thoroughly tested and evaluated before it can be considered fully operational. At a minimum, this includes testing cables. It can also mean analyzing network traffic prior to the company using the network; this can be most critical when usage is high.

feedback loop
Assessing an event and using that information to improve the process before the event is repeated.

NOTE

Between each of the phases, there is a **feedback loop**. A feedback loop is a way of taking what you have learned in one phase and using that information to refine the results of the previous phase. This ensures a better process in the end.

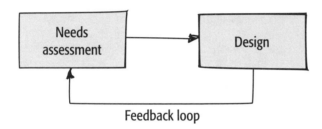

As with any design, there are many approaches to creating a good design and then a successful project. The steps described above provide a system to create a LAN design that will meet the most urgent needs of the users and the organization. So don't be surprised if you find that IT professionals are familiar with this system of LAN design. If they are experienced at designing they probably follow these steps and don't even know it.

Needs Assessment

The first step in LAN design requires that you do thorough research, what we call **needs assessment**. During the needs assessment phase, the focus is on collecting information about equipment, topology, network technology, and, in general, the needs of the company. Your task is to do the following:

- Inventory equipment currently in use
- Research current documentation like server log files and existing network diagrams
- Document the building layout and current cabling infrastructure if it doesn't already exist
- Determine what work the users need to accomplish by interviewing users
- Conduct meetings with key staff that can influence any phase of the LAN design
- Identify the functions and departments of the organization that rely heavily on the availability and health of the network

The needs assessment is a critical step in LAN design. Often, the needs assessment phase is rushed and therefore incomplete, which can lead to bad decisions and omissions. It can also lead to an increase in cost, sometimes a dramatic one that could have otherwise been avoided.

Needs assessment is a discovery process. Like an explorer in search of hidden treasure, your object is to find as much information as possible, analyze it, and make some decisions about what it all means. Unlike the seasoned explorer, you will probably experience some frustration the first few times you do a needs assessment. As your experience grows, you will become more thorough and efficient.

Equipment Inventory

An equipment **inventory** in a large organization can be the most time-consuming part of the needs assessment phase. If a company has never inventoried its equipment, or if the inventory is outdated or incomplete, you'll need to find and document every piece of equipment. Gathering this information is important for insurance and legal reasons, as well as for financial planning. That makes your job that much more valuable to the entire organization—not just to the IT department.

---

needs assessment
A process of collecting information about the needs of individuals and an organization to improve the performance of a system.

inventory
A spreadsheet or database listing details about the equipment owned or leased by a company.

LAN Design

When taking inventory, you must document important pieces of information about each machine. It is useful to create a database in which you can enter the information for each device; the database can then be used for reports and future additions. The reports can provide a quick analysis of the age of the computers and their capabilities. For performance reasons, you may decide to upgrade your computers based on this information. (You might also decide to get new computers if you have to upgrade to a new version of software that has minimum hardware requirements that the current systems do not meet.)

A couple of sample inventory sheets are shown as follows:

SAMPLE INVENTORY SHEET 1

COMPANY SERIAL #	MANUFACTURER SERIAL #	MAKE	MODEL	LOCATION	DESCRIPTION
10012	AJZ075DM	Dell	Opti-GX260	214	Desktop Computer
10013	DD7MF00	Dell	Prec-340	Reception	Tower Computer
10014	10007M-AF1178	Apple	G4	Senior Design	Tower Macintosh

SAMPLE INVENTORY SHEET 2

COMPANY SERIAL #	MANUFACTURER SERIAL #	MAKE	MODEL	HARD DRIVE	CPU	RAM
10012	AJZ075DM	Dell	Opti-G1	2.1GB	P 233MHz	32MB
10013	DD7MF00	Dell	Opti-Gn+	2.1GB	PII 300	64MB
10014	10007M-AF1178	Apple	G3	4GB	PPC 750	64MB

When completing the equipment inventory, it is important not to stop with the computers. Printers, hubs, bridges, routers, and other networking devices need to be inventoried with their corresponding network information. The information that you should collect includes:

- The MAC address
- The make and model of the NIC
- The IP address, if assigned
- Any technical specifications provided by the manufacturer
- The network type—10BaseT, 100BaseT, and so on

Collecting this information in an inventory can be very useful when managing the network later on. For example, the MAC address of a workstation or device can be used to identify the workstation in the event that there are duplicate IP addresses on the network.

Chapter 12

Hardware Identification

It's no surprise that finding your hardware information is difficult. Each type of operating system—whether it's Windows XP, Linux, or Mac OS X—has a place where this information is kept. You just have to know where to find it.

Identifying System Information in Windows XP When starting a PC, the first thing the computer loads before the operating system is software called the **BIOS**, which makes the computer function. The BIOS software loads right after you turn on the computer. If you press the F2 or Delete key when the computer first turns on (you usually see a prompt for which key and when to press it when the computer first starts), you should be in the BIOS setup page. Since the BIOS software is different for each computer, you will have to navigate through the software to find the CPU type and speed, the RAM, and the size of the hard drive.

Identifying System Information in Mac OS X When completing an inventory of a Macintosh, it is important to find the **Ethernet Address**. This address is found in the Network Csettings in the System Preferences. Click on the Built-in Ethernet option in the Show pull-down menu and click on the TCP/IP tab.

Other important information can be found in the System Profiler located in the Utilities folder, which should be located in the OS X Applications folder. The System Profiler can tell you the CPU type and speed, the amount of RAM, and other very specific hardware and software information.

BIOS (basic input/output system)
The BIOS service, located on a computer's ROM chip, enables the hardware and software to communicate with each other.

ethernet address
The term used to describe the MAC address on a Macintosh computer.

LAN Design

> **NOTE**
> Older versions of the Mac OS do not have the System Profiler application. Instead, you will need to use several different tools to get the necessary system information. Check the documentation that came with your system.

Documentation

One way for a company to save money is by documenting the technology infrastructure. Documentation should be kept on all aspects of technology in the company and should include:

- The planning process
- The network design
- Equipment
- Performance, troubleshooting, and system logs
- A blueprint of the network layout
- Meeting notes

The documentation provides valuable information about where problems are occurring and what design options are possible. A thorough documentation of the computer equipment can provide necessary details about the memory-processing and storage capabilities of the computers. This information becomes very valuable when considering integrating new applications that require more powerful computers. In the planning process, the documentation can make the difference between an unreliable network and one that will support a growing company for many years.

Another important part of documentation includes the network engineer's journal. The journal can be any print or electronic documentation tool that contains all information related to projects managed or completed. When collecting information about a problem, researching a solution, or brainstorming, the journal becomes the footprints that tell you where you have been and what you have done—and let you know in what direction you're headed. If nobody maintains accurate documentation, the history of the network is virtually lost.

So what happens to all this documentation over time? Well, it may appear to collect and sit dust, but eventually the documentation will be needed to solve a problem or explain past decisions. No one problem is more frustrating for a network engineer than creating documentation for a network designed and built by another network engineer. In an emergency situation, the absence of documentation can translate into hours or even days of downtime, which can result in an unacceptable and possibly debilitating end result.

Chapter 12

Facility Assessment and Documentation

In the second part of the needs assessment, you need to document the layout of the building. This includes identifying the size of the building and rooms, locating the offices, and noting the position of furniture. In addition, it's very important to include information on the location of electrical outlets, air vents, water pipes, and lighting. The easiest and quickest way to collect most of this information is to use a full set of blueprints, which will provide accurate information about the location and size of the facility. When completing a walk-through of the building, you should be looking for and documenting the following:

- Electrical outlets and wiring
- Phone cabling
- **Point of entry (POE)** for the telephone lines (also called the **demarcation** point or **demarc**)
- Ventilation shafts
- Lighting
- Location of water pipes

If there is existing network cable that is not included in the blueprints of the office or a physical network diagram, you will have to add it to your diagrams. Tracking the path that each cable travels can be a daunting task. But if you will be utilizing the existing cables, you'll need to know the path the cables take for troubleshooting and future planning purposes.

point of entry (POE)
The location in the building where the telephone lines enter from the street.

demarc (demarcation)
The location in your building or office where the local phone company's responsibility for the lines and equipment ends.

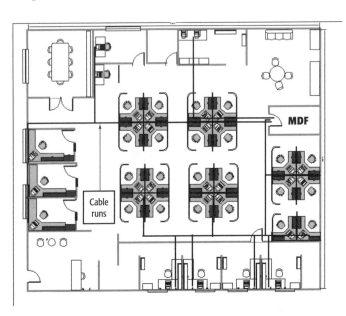

LAN Design

Assessing for Wireless Networks

Wireless networks pose a different design problem for a network engineer. Yes, wireless networks can replace some of your cabling infrastructure and reduce the challenges associated with moving people and computers around the office. But unlike a home office, in which a wireless network covers a small area, an office has many obstacles.

Designing networks to include wireless access means doing testing of the performance and reliability of the wireless technology before completing the design. Wireless testing involves the use of a **Wireless Access Point (WAP)** and a laptop configured with a wireless card. Using your test equipment, you will need to place the WAP in a wide-open location, such as a hallway. With the laptop, you need to walk around the office and test for areas where the connection is weak, unreliable, or nonexistent. During the setup for your wireless testing, you should be on the lookout for objects that may interfere with the performance of your wireless network. Some of these objects include:

- Solid concrete walls
- Large electric devices such as elevators and transformers
- Distances in excess of 500 feet (most WAPs work beyond 500 feet, but performance degrades quickly)

Wireless Access Point (WAP)
A network device that follows the IEEE 802.11a, 802.11b, or 802.11g protocols for providing network access to devices using radio frequencies.

termination
The location where a cable ends and connects to a jack or patch panel.

TIP
It is important to perform your wireless testing with the same brand and model of WAP and wireless card that you will be purchasing to use during the implementation.

Visual Inspections

One of several areas often overlooked in LAN design is the location of the cable **terminations** in the offices. You'll need to do a visual inspection to determine the best location for the network connectors, which can be located in the walls, floors, or partitions used in cubicles. Besides noting the proper location for the cable drop, it is important to talk to the people who work in the offices. They can provide information as to where furniture and other fixtures will be located in the future.

NOTE
It may not be cost-effective to draw a detailed diagram for each office, but it is critical that the visual inspection be completed.

Assessing User Needs

Assessing user needs is one of the most important parts of the needs assessment phase. You can conduct a survey of employees to better assess what the company will need. Although it may not be possible or realistic to interview everyone in the company, choose at least a few people from different departments and with different skill levels and responsibilities. The purpose of this interview is to determine the way people are using computers now—and how they are planning to use them in the future. Use the following questions as a starting point for interviewing users:

1. How do you currently use the network?
2. What information do you need to share with other employees?
3. What challenges or difficulties are you experiencing when trying to complete your work?
4. What applications do you use?

Communication   If your company is already using a network, it probably has messaging software such as electronic mail. If an e-mail server is in place, users may provide information about the performance and reliability of the system. There may also be a need for users to share their schedules and a contacts database.

TIP

If users are managing and maintaining their own contact databases and communications software, there may be an opportunity to streamline the system by centralizing services in a single product or solution.

Security   Security is a challenge for many small companies. Some users don't want any security, whereas others want you to take every security precaution possible. You need to consider password policies, physical access, and internal and external network security.

Support   One of the difficult aspects of technology is providing adequate and responsive technical support. Although the staff may primarily request workstation or software support, the problems they're experiencing may be due to poor network performance. Proper LAN design can greatly improve performance and reduce some of the support calls.

Creating the LAN Design

The information gathered from the needs assessment represents the completion of just the first phase in the process. Next, you need to analyze the information to determine the best design solution for your situation. The following sections cover the criteria used to make your selections. Your design will incorporate three main components:

- Network architecture
- Topology
- Network devices

Architecture Selection

As you learned in Chapter 3, there are several different network architectures, each of which is well suited to particular needs. Here is a brief review of each architecture:

Peer-to-Peer

- Designed for 10 users or fewer
- Allows users to share their files and printers with others
- Lets users manage passwords and access to resources
- Offers relatively easy setup and maintenance

Client-Server

- Supports large networks with thousands of users
- Provides a high level of security
- Features a dedicated server that centralizes resources
- Requires centralized management by a network administrator or team of network specialists

Hybrid

- Allows users to manage the resources on their computers, and gives network administrators the ability to manage the centralized resources
- Incorporates low and high levels of security

Chapter 12

The Architecture-Selection Worksheet

Knowing the characteristics of each of the network architectures makes selecting the right one for your company easier, but there are no steadfast rules. For example, an office of 10 employees may appear to be a perfect scenario for a peer-to-peer network. In actuality, the needs assessment may tell you that the company anticipates future growth, or that security is an issue and that people are concerned that their files are not being backed up. For this example, a client-server architecture would be the appropriate choice.

Architecture selection is a critical component that will drive much of the design decisions. If the architecture is peer-to-peer, there is less need for a dedicated wiring closet and high-speed connections. On the other hand, client-server architectures will require high-speed connections to the server and a secure room to house the cables and the server or servers.

To determine the right architecture, you can create an architecture-selection worksheet, in which you can rate the importance of things such as performance, security, and centralization; and you can also note what you can spend on the network, as well as whether or not a full-time position can be dedicated to managing the network.

Using such a worksheet and what you know about each of the architectures, you can make an informed decision. For example, a company of 10 employees would qualify for a peer-to-peer network, just based on the number of employees. But if they gave a high rating to security, they probably would need either a client-server or hybrid architecture. A high score for performance would not help you decide on an architecture because it is possible to get pretty good performance from a peer-to-peer architecture. If a full-time management position is possible, then you could exclude the hybrid option. If there are dedicated network management personnel, there is less of a need to complicate things by using a hybrid model. (The only exception would be if the employees specifically stated they had a need to share resources on their computers.)

NOTE

Many businesses that have dedicated network staff implement only client-server architectures. The reason for this is that the network staff can keep very tight control on security, minimize desktop maintenance, and lower support costs by preventing users from adding applications or changing the workstation settings. Creating generic workstation settings with the same applications is called *standardization*.

268

LAN Design

Topology Selection

The next step in the design process is to identify the topology to be used. This includes both the logical and physical topologies discussed in Chapter 5. Although two logical topologies, ring and bus, are possible, this section will assume the more predominant bus topology found in Ethernet networks.

As you have learned, the original Ethernet networks used various types of coaxial cable. In the early Ethernet networks, the coaxial cable was strung along from computer to computer in a **daisy-chain** fashion. Modern Ethernet networks follow a star pattern, or more accurately, an extended star pattern. In an extended star topology, there is a core hub or switch that has cables radiating out to other hubs. The peripheral hubs have cables that radiate out to the computers or other network devices.

daisy chain
Using a cable to connect devices together from one to the next so that the output of one device is connected to the input of another to form a chain.

distribution layer
The devices that provide access to the network backbone or to the core layer.

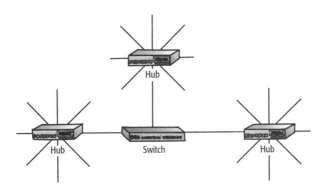

access layer
The layer in which network devices act as an intermediary connection point between the end systems and the distribution layer.

The core hubs are organized into a group called the **distribution layer**. The distribution layer contains the most powerful hubs that are used to support the backbone of the network. In some large installations, there may even be a core layer of switches and routers that interconnect multiple LANs using powerful, high-speed network equipment. The hubs that connect to the distribution layer make up the **access layer**. These hubs provide access to the network to end systems such as computers and printers. Access-layer hubs can vary from the simplest devices to very fast and sophisticated ones, depending on the need and the costs.

main distribution facility (MDF)
The central wiring closet that is used to house the core network devices such as the core hub, switch, and routers. All other wiring closets connect directly to the MDF.

The Wiring Closet

In every network installation, there needs to be at least one room where all of the cables come into. This closet is called the **main distribution facility (MDF)**. The MDF contains the distribution-layer devices—the switches or hubs—as well as any access-layer devices. The MDF may also be used to house the telephone

269

system and even the file servers. If the office is large or inhabits a multistory building, you will need to have **intermediate distribution facilities (IDFs)** on each floor. Each IDF is directly connected to the main hub or switch in the MDF. The backbone connections between the MDF and IDFs are usually fiber-optic cable, although copper is sometimes used as a cost-saving measure.

> **intermediate distribution facility (IDF)**
> A wiring closet that is an extension of the MDF; used to reach devices that are too far from the MDF.

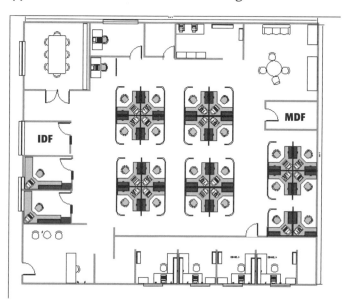

Sizing Up Your Building

Depending on the media you are using, there is a helpful way to determine whether you will need more than one wiring closet on each floor. On your blueprints, locate the room that you want to use for the MDF. Assuming you are using Category 5e cable, use the MDF as your stating point. Draw a circle around the MDF with a radius equivalent to 65 meters. If the offices are all 65 meters or closer to the MDF, there is no need for an IDF. The 65 meters leave enough extra cable to account for corners, going up and down walls, and for the patch cables.

Selecting the MDF Location

Since the MDF will house much of the sensitive equipment for your network, it is important that you select the location of the MDF carefully. Selecting the location of the MDF involves finding a room that meets several criteria:

Room Size   Depending on the size of the company and the number of cable drops, you may need a room as small as four feet by six feet or possibly at least 10 feet by 10 feet. Plan a room that provides adequate space for the network

LAN Design

staff to easily move around the equipment and do whatever work they may have to do in the closet.

Power There needs to be enough power for all of the equipment using multiple electrical circuits. Check with a licensed electrician.

Lighting In any of the wiring closets, it is important to have adequate lighting. Although florescent lighting is acceptable, it needs to be far enough away from the cable and equipment to protect against the electromagnetic interference generated by the fluorescent lights.

Ventilation The equipment in wiring closets generates a lot of heat. Since the equipment is very sensitive to heat, it is important that the wiring closet have adequate ventilation. There are two ways to accomplish this: One is to install a ceiling or wall fan that is either on at all times or is triggered on by a thermostat. The second option is to install air conditioning or provide ventilation from the office air conditioning. The MDF will need its own thermostat.

Security The location of the wiring closet presents a major security concern. You want to design a network that limits access to the wiring closet to only the appropriate people. A wiring closet should be in a secure room with a door and a lock with its own key code.

> **patch panel**
> Also referred to as the *horizontal cross-connect*, the patch panel is mounted on a rack and is used to connect the cables from the wall outlets to the hubs in the wiring closets.

WARNING
The entrance to the MDF should require a key to get into, but not require a key to get out. Check your local fire and building codes for applicable laws.

Plumbing Moisture of any kind can seriously damage network equipment. The ideal wiring closet will be free of any water sources in the ceiling, floor, and walls. This will protect the equipment in the event that a water pipe breaks. If your area is prone to flooding, you will want to consider putting your equipment on the second floor. Sprinklers are another danger. There are special fire extinguisher systems that use chemicals instead of water, but they tend to be expensive.

Planning the Cable Runs

After you have identified the locations of the MDF and any IDFs, you need to plan the cable runs. Following EIA/TIA standards, each cable run should not exceed 90 meters. When measuring, this will be the distance from the **patch panel**, also referred to as the *horizontal cross-connect*, in the MDF to the jack located in the wall of the office. (This cable run is called the *horizontal run* because it traverses the floor of the office building.)

Chapter 12

The next step requires using the building blueprints. During the needs assessment phase of the project, you should have verified room numbers and locations. Using this information, you can create a cut sheet. The cut sheet lists the cable IDs, the location of the cable terminations, and cable length:

Time Domain Reflectometry (TDR)

A technology found in cable testers, TDR sends a signal down a copper wire and waits for the signal to bounce back. Based on the amount of time it takes for the signal to return, the TDR cable tester can determine the length of the cable.

Sample Cut Sheet

CABLE ID	LOCATION	TERMINATION	LENGTH
100-A	RECEPTION	143-MDF	87 METERS
101-A	ROOM 101	143-MDF	72 METERS
101-B	ROOM 101	143-MDF	72 METERS
202-A	ROOM 202	226-IDF	64 METERS
210-A	ROOM 210	226-IDF	45 METERS

Although you will have an approximate idea of the length of the cable run, you should document the exact length after the cable has been pulled and terminated. Any cable tester that uses **Time Domain Reflectometry (TDR)** is capable of calculating the length of the cable.

Drawing Your Network

An important part of LAN design is drawing diagrams of your network. Drawing tools such as Visio from Microsoft Corporation allow you to create detailed drawings, right down to the make and model of the equipment. Visio Enterprise has a new feature that will auto-discover devices on your network and attempt to identify the type of device and then create the logical drawings of the network. This feature can reduce your drawing time significantly.

When you create network drawings, it is always helpful to have the blueprints for the building. Better yet is to ask for the electronic versions of the blueprints. Architects will typically use an application called AutoCAD to create blueprints. The AutoCAD program saves the files in the .dxf format. Using Visio Professional or Enterprise, you can open the AutoCAD files and draw your network directly into the blueprints. For more information about AutoCAD, go to www.autodesk.com.

TIP

If you are planning a new LAN or upgrading an existing LAN, you should plan on purchasing only switches. Since the price of switches has dropped dramatically in the past three years, hubs (discussed in Chapter 9) are no longer cost-effective.

LAN Design

Device Selection

Now that the wiring closet and cable runs have been identified, it is time to take a look at the devices that will make the network operational. The main components you need to select include:

- Switches
- Routers
- Wireless Access Points

Choosing Switches

A switch improves network performance by localizing traffic on each port. Traffic on one port is not transmitted out to all ports, as happens in hubs. If a source device wants to communicate with a device on another port, the switch will forward the frame.

Like hubs, switches are also available in 10Mbps, 100Mbps, and 10/100Mbps configurations. More recently, many companies, including Cisco, have begun selling gigabit switches. Gigabit switches operate at 1000Mbps, and are best suited to backbone applications at the core of a large enterprise network.

The reasons to choose switches are:

- They improve performance on the backbone.
- They function as distribution-layer devices that are at the center of the network and have hubs extending from them.
- They can also function as access-layer devices to improve performance to the desktop with dedicated switch ports for each computer.
- They minimize collisions on the network.
- They provide increased security.
- They allow you to monitor and control traffic using SNMP functionality.

NOTE

Although you may have the good intention of buying equipment from several different companies, beware of compatibility problems that typically arise in more complex LANs using switches with advanced features. If you are considering migrating to switches, you may want to purchase all network infrastructure equipment from the same vendor.

Choosing Routers

The addition of switches to your network immediately makes the LAN a reality. One device remains to be added to complete your network. A router provides two critical functions:

- Provides network segmentation in situations such as multistory buildings or large networks, which will improve performance and management
- Provides access to other company networks located at other sites and networks found on the Internet

Most companies require at least Internet access, if not segmentation as well. Selecting the right router requires an understanding of the needs of the company and the capability of the router. A small office with moderate bandwidth requirements such as DSL would be quite happy with a low-cost router; on the other hand, an Internet startup company that requires high-speed access to the Internet would need a powerful router, such as those provided by Cisco Systems or similar router companies. The differences between small office routers and more expensive models are performance and features.

Choosing Wireless Access Points

Like most network devices, Wireless Access Points do not perform or function the same, even though they are all designed to meet the same wireless standard. Wireless Access Points ideally have the following features:

- Minimum range of at least 1000 feet
- Support for 802.11b and the capability to upgrade to 802.11g
- Dual-band mode that supports connectivity for both 802.11b and 802.11g devices
- Web-based management for configuring, troubleshooting, and managing connected users

The wireless needs for each organization very greatly. Be sure to plan to have enough WAPs for your network. The most common standard, 802.11b, operates at 11Mbps. That limits the total number of simultaneous wireless devices to probably 20 or fewer. More costly WAPs typically perform better under a heavier load. If you are installing the WAP in a small office environment, a less-expensive model with a shorter range and support for only 802.11b may meet all your organization's requirements.

LAN Design

Installation

Once the design is completed, it will need to go through several revisions. After it is thoroughly developed, it can finally be implemented. The implementation phase marks the beginning of the actual installation of the network infrastructure. Up to this point, everything has just been on paper.

In preparation for the installation, a budget will need to be developed that will include costs for equipment, installation, and testing. If the design is created in-house, it will save the company money. If not, the costs for design will need to be included in the budget.

Media Selection and Installation

As you learned in Chapter 8, there are several different types of network media available. The most common types are copper and fiber-optic cables. The majority of networks installed today use Category 5e cable to the desktop. In some cases, Category 6 is used for future bandwidth needs and multimode fiber-optic cable is used primarily for backbone network installations or connections to powerful servers.

Category 5e twisted-pair cable is very inexpensive. When planning for the installation of Category 5e cable, take into consideration that most of the cost of installing the cable will be for the labor and not the materials. At a minimum, every location should have at least two cables per office.

When installing Category 5e twisted-pair cable, you should do the following:

- Determine whether **PVC** or **plenum** will be necessary.
- Decide whether or not **conduit** will be necessary or possible.
- Confirm that all cable runs are 90 meters in length or less, and that the total amount of cable from the computer to the switch does not exceed 100 meters.
- Determine whether telephone and video cable will be installed along with the Category 5e cable.
- Determine whether you'll be terminating the cable to a patch panel or a **punch block**.

Fiber-optic cable has become a common part of networks in businesses, government, and education. Fiber-optic cable comes in two types: single mode and multimode. Multimode is the predominant cable type. Fiber-optic cable comes with 2, 4, 6, 12, or even 36 optical fibers per cable. The fiber-optic cable needs to be installed by a professional cable installer who has experience working with fiber optics.

PVC
The plastic outer housing typical of data cable.

plenum
The type of outer housing of a cable that does not emit toxins when it burns.

conduit
Metal or plastic pipe or raceway used to protect the cable runs from physical damage or electrical interference.

punch block
Similar to a patch panel, except that it mounts directly on plywood on the wall in the wiring closet. Punch blocks are typical of telephone cable installations.

When installing fiber-optic cable, you should do the following:

- Verify that the contractor has installed fiber before; ask for certifications from Lucent Technologies or other similar leaders in fiber-optic networks.
- Determine if your equipment will require ST, SC, or another type of fiber-optic connector.
- Make sure that the fiber-optic raceway (the pipe that holds the cable) has the appropriate radius bend (a minimum of 90 degrees is recommended) at all corners.
- Make sure the fiber-optic cable is well-marked (building codes in some areas require warning signs indicating the presence of laser light).
- Do not place fiber-optic cable in the same bundles as Category 5e cable; it can be crushed if it ends up at the bottom of the pile.

Testing and Certifying Cable

Even the most thorough visual inspection of a cable installation can't ensure that all cables will function at their peak. Testing the cables is the only way to ensure that they will work. There are many different types of cable testers. Some inexpensive testers perform only **continuity tests**. Continuity tests only inform you if the cable can transmit an electrical signal. Other more expensive testers use a more sophisticated technology called Time Domain Reflectometry (TDR). If you are considering purchasing a tester, we recommend that you invest in a tester that uses TDR. More advanced testers that use TDR can give you valuable information, such as the following:

- Cable length
- Location of possible breaks
- Level of **attenuation**
- Amount of **near-end cross talk**
- Level of noise
- Whether or not a cable meets a specific category rating (e.g. Category 5e)

If a cable contractor is performing the installation, it is important to request a printout of each cable installation. The documented installation provides two valuable pieces of information: For one, it gives you a written record of the best possible network performance that you can expect on a cable. For example, if a cable is able to transmit at only 87MHz, it will never provide the 100Mbps throughput that is expected in 100BaseT. Secondly, it requires that the cable contractor verify that they have indeed certified all the cables to 100MHz, the

---

continuity test
A pass-or-fail test used to determine if an electrical circuit exists between a pair of wires.

attenuation
The degradation of a signal because energy is lost as it moves down the wire.

near-end cross talk
Interference on a wire from the electromagnetic field of another wire that was terminated incorrectly.

LAN Design

optimal frequency for Category 5e networks. You can request the Category 5e certification on all cables. This will add an additional cost to your budget, but it will mean a lot more in terms of performance in the future.

Connecting the LAN Devices

After the cable contractor has completed the installation of the network cable, the network administrator will be responsible for installing the LAN equipment. The first step after taking inventory of the equipment is to make sure that it is placed in the right wiring closet. This can make an important difference in the performance of the network. A misplaced high-end 100BaseT switch can end up providing the marketing department with great performance for printing, while the rest of the ordering-processing department sit and wait for their orders to be processed on the misplaced low-end switch.

The switches need to be securely mounted in a rack of some of kind. There are freestanding racks that are bolted to the floor; wall-mount racks that are bolted to the wall; and cabinet enclosures, which are the most expensive, but provide a completely enclosed solution that that can be moved when placed on casters.

When mounting the LAN devices, make sure to leave some space for cable-management panels. Cable-management panels provide a way of organizing and protecting cables as they wind their way from the patch panel to the switch. It is important that the **patch cable** for the server be connected into the correct port on the switch.

The following illustration depicts a typical layout of an equipment rack:

patch cable
As specified by the EIA/TIA 568 standard, a network cable that is used to connect an end system to a cable that is terminated in the wall.

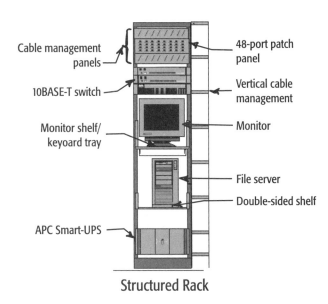

Structured Rack

Connecting to a WAN

Connecting to a wide area network through an Internet service provider requires a router on the company network. That is not all. Before the LAN is installed and operating, the network administrator is responsible for identifying a vendor to provide a WAN link for Internet access and other possible uses. In addition to finding a competitively priced ISP, you need to contact the telephone company for a dedicated line. This may come in the form of a leased line that provides digital service or a regular phone line for access via a modem or DSL. Often, the Internet service provider that is chosen will quote a price for the total charge of the WAN connection. Once an agreement is made with your company, the ISP will request the installation of the dedicated line from the telephone company.

firewall
Devices used with specialized software to secure a LAN from access by unauthorized sources.

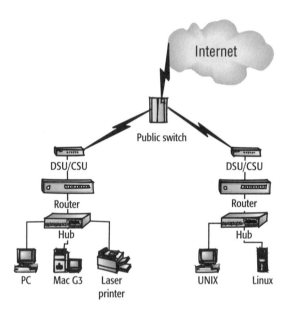

Depending on the type of service that is selected, it may be necessary to purchase additional equipment. Services such as DSL, Frame Relay, and PPP connections all require special equipment to transmit over the telephone company's lines. The same is true for accessing the Internet over cable connections, although cable Internet access is typically associated with residential access. (You'll learn about all of these services in Chapter 14.)

Before installing a router for WAN access, you should also make sure to discuss security issues with the Internet service provider. Cisco routers can be purchased with software that will provide **firewall** features. The firewall can protect your network from intruders poking their noses around your network and possibly doing damage.

Review Questions

1. What are the phases of a successful LAN design?

2. What information do you document during a needs assessment?

3. Why is an equipment inventory important?

4. What information needs to be documented in a facility assessment?

5. In what situation would you choose a client-server architecture over a hybrid architecture?

6. How do you determine whether an IDF will be needed in addition to an MDF?

7. What is one obstacle to assess when including wireless networks in your LAN design?

8. What is the difference between a distribution-layer device and an access-layer device?

9. What important functions does a TDR cable tester provide?

10. What are two purposes of using a router on your LAN?

11. When coordinating Internet access, whom does the network administrator need to contact?

Terms to Know
- access layer
- attenuation
- BIOS
- conduit
- continuity test
- daisy-chain
- demarc
- demarcation
- distribution layer
- Ethernet Address
- feedback loop
- firewall
- intermediate distribution facility (IDF)
- inventory
- main distribution facility (MDF)
- near-end cross talk
- needs assessment
- patch cable
- patch panel
- plenum
- Point of entry (POE)
- punch block
- PVC
- termination
- Time Domain Reflectometry (TDR)
- Wireless Access Point (WAP)

Chapter 13

Network Management

Building your first network may seem like a daunting task, but it is manageable given good planning and time. When you make it through building your first network—and you will—or rebuilding an existing network, you are going to need to think about what devices you are going to connect to the network. At that moment, you will have your first opportunity to make some decisions about network management.

Network management involves a network administrator performing a combination of several tasks. The tasks may vary, and the execution of those tasks may occur at different intervals, such as daily or monthly. The types of tasks involved in network management include monitoring network performance, ensuring security, maintaining fault tolerance, and troubleshooting. Network management begins with the planning of a network and continues on for the life of the organization.

In this chapter, you will learn:

 The importance of network management

 How to provide basic support with data backups, reliable power, and infrastructure redundancy

 Performance-monitoring techniques and tools

 How to monitor your network with the network management system and network management protocol

 A layered approach to troubleshooting networks

Chapter 13

Why Network Management?

As a network administrator, you will have the responsibility of ensuring uninterrupted network service to users. Having great troubleshooting skills may be a great asset for a desktop support person, but it is not adequate for a network administrator. You need to know about problems before they exist. How is this possible? Solid planning, the right monitoring tools, and detailed documentation will help you stay a step ahead.

Why is network management important? Network management represents the culmination of the past, current, and future work on a network. In essence, network management is the glue that ensures that legacy systems continue to operate with existing and new systems. And in the end, this glue should be transparent to users in your organization.

Networks are increasingly complex entities that change minute by minute. The challenge for a network administrator or information technology team is to have all the tools necessary to identify those changes and determine whether a change is critical enough to warrant a reaction. It may be that a device on your network, such as a file server, is operating near full capacity. If the file server is under heavy load only in the morning when everyone is logging in, it may be determined that it is safe to ignore this new information. But if the server is under heavy load most of the time, then you would need to take a closer look.

You can think of the tasks involved in network management as three points of a triangle. Each point is critical to the successful operation of the network. The three points include:

- Performance monitoring
- Technical and end-user support
- Documentation

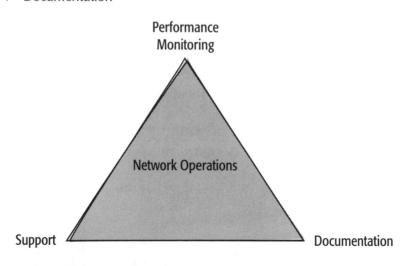

Network Management

If any of the three points are missing or incomplete, the network is directly impacted. Although each point might be regarded as a separate task in network management, the triangle makes it clear that each point is dependent on the other to maintain a healthy network.

Well-planned and executed network-management procedures and policies can help reduce the total costs of operating a business. Networks can perform several roles, one of which is the automation of a process that was once done by hand. Printing envelopes—a task that once would have taken several days—takes only a few minutes with a database of addresses and a functioning network. But if the network is down, it could take hours or days to retrain staff or reorganize a department to use a manual system. Network **downtime** is most often the reason that a company loses money. In many cases, solid network management can prevent downtime.

When a network management strategy is well-planned and implemented, it is critical that information on the health of the network can be analyzed. **Latency** and **bottlenecks** can be identified as they occur, rather than after several users report the problem. Several tools are available to collect data on network traffic, hardware performance, system performance, and security on devices. As you read through this chapter, you will be introduced to a few tools that are readily available for you to download and install.

Part of network support includes using tools to stay abreast of changes on the network. It also includes keeping users informed about their responsibilities in maintaining a healthy network. Many times, users don't understand everything that goes into building and maintaining a network. Although they typically don't need to know the gory details, it's helpful to explain to users networking basics such as accessing the server, software errors, and the backup process.

In addition, technical support needs to be responsive, reliable, and predictable. Set guidelines for response times—you may find, for example, that a four-hour response works great for general questions. One approach adopted by some IT departments is to provide support in levels, depending on the problem.

downtime
The amount of time that a service, such as e-mail or a file server, is not functioning.

latency
A delay in the transmission of data usually caused by excessive traffic.

bottleneck
A device on the network that is slow in transmitting data, causing delays in the delivery of data.

```
End-User Support Resources

Level 1: Desktop Support

Level 2: Diagnostic and Network Support

Level 3: Vendor or Software Engineering Support
```

Chapter 13

The first level of support, End-User Support Resources, involves creating documentation for your users. This can include resources from vendors, how-to articles, and a list of remedies for common problems called a **FAQ (frequently asked questions)**.

FAQ (frequently asked questions)
A list of specific questions about the most common problems encountered by users, followed by the answers.

> **TIP**
>
> Training has purposely been left out of the support model because support indicates a reactive approach. Depending on the organization, training may be handled by the human resources department, the department the user works in, or the Information Technology group. Remember, even the best technical support is not a cost-effective replacement for quality technical training.

If the user is unable to resolve the problem on their own, they will request the front-line services of the Level 1 support team. Level 1 support personnel are responsible for resolving desktop application issues, assisting with software installation or configuration, and troubleshooting network connectivity. Typically, Level 1 support utilizes e-mail or phone to provide a cost-effective support model. More recently, remote desktop control applications allow Level 1 support to view the steps performed by the user or take control of the user's computer when necessary. If all else fails, Level 1 support staff may be called on to provide on-site support, although this is provided as a last resort.

Level 2 support staff are more skilled and experienced in troubleshooting. These folks also have more in-depth knowledge of desktops, networks, and specialized network devices such as firewalls and routers. Unlike the technical support provided by a specialized software or hardware manufacturer, the internal support staff have to know how to deal with dozens of software applications and hardware in a myriad of scenarios because the cause of a technical problem can be very different given a different network scenario.

When all internal support resources are exhausted, the IT staff turn to outside resources for help. Vendors may be contacted for specific troubleshooting support or to report an undocumented software problem. Or, in the case of customized applications, the software engineering group will need to get involved to determine whether changes need to be made directly to the software. If the latter is true in this stage of the support model, expect to have an alternative solution ready for your users because it can take months before the software engineers or vendors are able to provide a fix.

Finally, network management would not be complete without up-to-date, accurate, and complete documentation. You learned about part of the documentation process in Chapter 12. Documentation also includes reports on network performance and security, log files, and administrator journal entries.

Network Management

First Steps in Network Management

Whether you are currently managing a network or are taking over a network from another administrator, it is very important that the basic components for protecting and managing the network are in place. Initially, you will need to evaluate some basic systems that will ensure that there will be no loss of data in the event of a hardware failure or catastrophe such as a fire or natural disaster. Although you have the goal of providing continuous, uninterrupted service to users, reaching that goal won't mean much if some or all of the user data is lost.

Two key concepts define the first objectives that a network administrator should achieve: **Fault tolerance** is found in networks that are able to withstand a partial failure and continue to operate, albeit with some impact on performance. Ensuring the highest levels of fault tolerance requires **redundancy**. Redundancy, which can occur at many different places, is used to provide an exact duplicate of the primary system in hardware, software, and data. You will recall from the discussion of Windows 2000 Active Directory in Chapter 4 that multiple domain controllers are used to keep duplicates of the Active Directory database.

Although large organizations will require more complex solutions, at a minimum you will need to consider three systems that need to be in place to support your network infrastructure:

- Tape backup system
- Uninterruptible power supply
- Redundant servers and server hardware

These topics were briefly covered in Chapter 4, and we will review them in more detail here.

fault tolerance
The use of hardware and software to prevent the loss of data and downtime.

redundancy
A method of providing fault tolerance by duplicating a service or hardware.

TIP

Most companies who sell network servers offer tape drives as added components. Although internal tape drives are an option, external tape drives are preferred because they can be powered on and off without powering off the server. If maintenance is required, an external tape drive can be replaced by a backup or new drive and sent to the manufacturer.

Chapter 13

Backing Up Data

A backup is a process of copying data stored on a computer and saving an exact duplicate of the data on another storage device. Backups of data can include operating systems, user files, applications, and just about anything that is stored on a hard disk. Backups can be as simple as copying a document onto a floppy disk for storage; or they can involve special backup software, hardware, and media.

Typically, backups of servers and computers that maintain important data are performed on dedicated hardware using special software. Tape drives are used to store large amounts of information at a relatively low cost. Most drives use **SCSI** communications, although high-end systems use fiber-optic connections for the best performance when transferring data. There are several types of tape drives:

Travan   Travan tape technology was developed by Imation. A low-cost solution, Travan tape drives are hampered by low capacity and very slow transfer rates, but are considered an an ideal solution for small servers that have fewer than 20GB of data to back up.

DAT (Digital Audio Tape)   DAT was developed by Sony for recording digital information. Originally developed for audio, DAT drives have proven very effective for storing computer data. Two standards exist for DAT drives: DDS-2 drives are capable of storing 4GB of data in uncompressed mode and 8GB of data in compressed mode; DDS-3 drives cost more, but they have a storage potential of 12GB of data in uncompressed mode and 24GB data in compressed mode.

AIT (Advanced Intelligent Tape)   AIT was developed as a standard by the music recording industry. AIT drives are capable of storing large amounts of data—as much as 50GB per tape—and have transfer rates up to three times as fast as DAT drives. AIT tapes use a special chip called a **MIC** that is located on the tape media. The MIC stores information about the contents of the data on the tape. (DAT and other types of drives usually record this information directly on the tape media.) The MIC chip on AIT tapes can cut the backup time in half.

DLT (Digital Linear Tape)   DLT was developed by DEC more than a decade ago as a reliable, high-speed backup solution for mainframes. In the enterprise market, DLT is considered more reliable than DAT or Travan technology although it has been overshadowed by the new Ultrium (LTO) and SuperDLT technologies that provide faster performances and much larger backup capacities. DLT is now a licensed technology from Quantum.

Ultrium (LTO)   Ultrium is one of the most recent and fastest tape solutions. Ultrium is the implementation of an open tape technology called Linear Tape-Open (LTO), which was jointly developed by IBM, HP, and Seagate. The Ultrium format provides a reliable, fast, and large-capacity solution. Each Ultrium tape is capable of storing 200GB of data—assuming a 2:1 compression ratio and transferring at rates of 40MBps. That equates to 2.4GB per minute.

SCSI (Small Computer System Interface)
A high-speed interface for connecting hard disks, CD-ROMs, and other peripherals to a computer.

MIC (memory in cassette)
A flash memory chip imbedded in an AIT cassette that stores the catalog of the tape's contents.

Digital Linear Tape (DLT)
Developed by DEC as a reliable, high-speed backup solution for mainframes, DLT technology has been considered one of the most stable and widely proven tape back technologies..

Ultrium (LTO)
Ultrium is the implementation of a reliable, fast, and large capacity open tape technology called Linear Tape-Open (LTO) ,which was jointly developed by IBM, HP, and Seagate.

Network Management

SuperDLT Like LTO technology, **SuperDLT** represents the modernization of the DLT format, taking advantage of new optical and magnetic technologies that can place more tracks of data on the tape media. SuperDLT is a new technology from Quantum. Now in its second generation, SuperDLT currently supports storage capacities of 320GB on one drive using 2:1 compression, and transfer rates of more than 32MBps—assuming the same compression ratio.

The speed of the drive is an important factor when performing tape backups. Completing a tape backup of 8GB of data can take as long as four hours. In situations where the storage need is much greater, **autoloader** tape drives can be used. Autoloader drives come in varied configurations. An Ultrium tape autoloader drive with nine tapes can record more than 1TB of data. When purchasing tape drives, check with the manufacturer about recommended media. The quality of the tape can have a significant impact on the reliability of a backup. Also, be sure to review the cleaning requirements for the tape drive mechanism because this can vary greatly between tape backup technologies.

Why Do Backups?

Tape backups are a critical part of network management. Regardless of the size of the company, whether it is a one-person home office or a multiserver network, backups can become the lifeblood in the event of a disaster. Backups serve several important functions:

- They provide an inexpensive storage option.
- The portability of tapes allows for off-site storage.
- Large amounts of data can be backed up at once.
- Recovery from a crash takes hours, not weeks.
- They save time.
- If a server crashes, the entire system can be restored on a replacement server.
- Files that are accidentally erased can be individually restored.

SuperDLT
The modernization of the DLT format taking advantage of new optical and magnetic technologies that can place more tracks of data on the tape media. See also *Digital Linear Tape (DLT)*.

autoloader
A multidrive tape device that holds several tapes and automatically switches to the next tape when a tape is filled.

TIP

Although the whole purpose of a tape backup is to be able to restore the data when necessary, you may find yourself in a situation in which the restore does not work. The best way to avoid this problem is to replace the tapes regularly (check with the manufacturer), clean the tape drive regularly, and make sure you have access to another tape drive that is the same model as your backup drive for situations in which the drive fails.

There is no reason not to do a tape backup, no matter what the situation. You could lose your data for any number of reasons: A virus could erase your hard drive; the hard drive could fail; a new installation of software could render your computer useless. Or you could just accidentally erase a critical file. Furthermore, it is realistic and prudent to consider major catastrophes as potential risks for data loss. Offsite storage is the right solution for this problem. Fortunately, there are many companies that will provide pick-up, storage, and drop-off of your backup media for a nominal fee.

> **archive bit**
> A bit setting for a file that is used by backup programs to determine when the file was last backed up. The bit is reset after every differential and normal backup.

Types of Backups

There are different types of tape backup that vary in the amount of time it takes for the backup process to be carried out. Depending on your needs, you can use differential, incremental, or normal backups. If you use a differential or incremental backup, the backup software will scan files that have their **archive bits** turned on. This means that the backup software assumes that the file has changed since the last backup.

Normal   Normal backups are also referred to as *full backups*. Normal backups are used to back up all data, whether or not all the files have changed since the last backup. This type of backup is used the first time a backup is performed on a server and then usually once a week to back up all data.

Differential   Differential backups use the archive bit to determine whether a file has changed since the last normal backup. Differential backups take longer than incremental backups, but with them it takes less time to restore data. Use this type of backup in situations in which it is important that the restoration be completed quickly.

Incremental   Incremental backups use the archive bit to determine whether a file has changed since the last incremental backup. Incremental backups take less time than differential backups. However, data restoration takes more time because the normal backup tape and all incremental tapes made since the last normal backup are needed. If a normal backup was completed on Friday, and the backups made on Monday through Thursday of the following week were incremental, you would need to use all five tapes to perform a complete restore.

When planning your tape backup strategy, the normal backup usually takes place on Friday evenings. You can set the tape backup to occur after everyone has left for the day. Because normal backups record all data, the backup can take a long time to complete. In companies that require support over the weekend, the tape backup strategy needs to be carefully planned so that all data are backed up but the operation doesn't interfere with the normal functions of the company.

Network Management

Planning a Backup

Before performing a tape backup, the network administrator needs to decide on a backup strategy. The backup strategy is a decision about how far back the tapes need to be kept before they can be used. Is it one week, two weeks, or four weeks? Depending on your decision, you will need to have a tape in the set for every day.

If you don't have much data to back up, and you are planning on doing a full backup every evening, you might need just one tape. This strategy is simple, but it doesn't provide adequate protection against data loss. If a different tape is used for each day, the strategy is more reliable, but not as practical. Performing a normal backup every evening is time-consuming. Also, the majority of the data being copied doesn't change from day to day, so duplicate information is being stored from the day before.

The most common strategy involves a rotation scheme. Four tapes are used Monday through Thursday for either differential or incremental backups. A separate tape is used for each Friday of the month. On the last Friday of the month, the fourth Friday, the tape is kept as an archive. The total number of tapes needed for this strategy is 19.

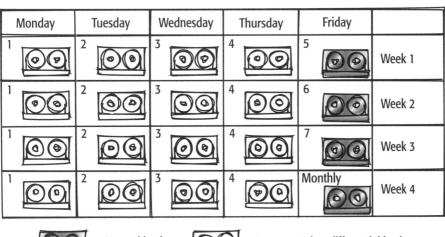

There are many backup programs on the market. Most products—including Seagate Backup Exec, Computer Associate's ArcServeIT, and Dantz's Retrospect—allow multiple servers to be backed up on a single tape drive. Versions of UNIX, Windows 2000, and Novell NetWare come with limited backup software. Evaluate your needs carefully to determine which software is right for your network.

Uninterrupted Power

In addition to implementing a backup strategy for your data, you need to protect your sensitive electrical equipment from damage caused by brownouts, blackouts, sags, surges, and spikes. It may take only one hit from a bolt of lightning to irreversibly damage equipment.

Uninterruptible power supplies (UPSs) are battery backups that supplement the current that is usually available from the wall outlet with current supplied by a battery. You can evaluate your power needs by asking yourself some basic questions:

What devices need to stay up and running? In most cases, only the servers will have an uninterruptible power supply (UPS). If the information running on workstations is critical, it is important to provide battery backup to them as well. Workstations that do not have backup power will lose all data in memory if the power goes out. Also, the sudden loss of electrical power can permanently damage data and the hard drive.

How long does the network need to be running after a power outage? If the network needs to keep running for more than a few hours, you need to consider either a very large battery system or an expensive generator.

How much power does the UPS need to support to maintain all of the equipment for the total amount of time required? Check with the manufacturers of UPS systems to determine your battery backup needs.

The UPS can provide many functions beyond the role of temporary power source. Most of the UPS systems on the market offer more than just battery backup for power failures. They can provide line conditioning, which can be used to supplement power during brownouts and sags. In addition, the UPS can also absorb spikes and surges, in which case there's no need for surge protectors on the line as well. There is also software available that will interact with the UPS and servers. It will shut down the servers if the power doesn't come back up in time and the UPS is starting to run out of battery power.

TIP

If power goes out at your site, it is possible that there's also no power at the CO of the telephone company. Companies that rely completely on their Internet connection for business, such as online trading companies, should contact the telephone company to determine how long the telephone company will stay up and running.

Network Management

Redundancy

Redundancy is a system of duplicating a service or function that already exists on a network. Redundancy can be applied almost anywhere in a network, from hard disks to network cables. More recently, the dependence on Internet access for business transactions and communication has prompted many companies to add redundant Internet connections. Most of these companies use two separate providers for more reliable service.

Redundancy is used to provide continuous uninterrupted service to users on the network. Tape backups and UPS systems do not provide redundancy, but you can add a redundant tape backup and UPS in the event the primary units fail.

When to Use Redundancy

There are many situations in which you should use redundancy:

- You should add redundancy to the file server first, in the form of a RAID solution.
- You should add redundant features to any component that must always be running.
- You should use redundancy for business-critical functions such as a database server or Internet access that clients depend on.
- If you would lose your job if the network goes down, then you need to add redundancy wherever possible.

Implementing Redundancy

If money weren't a problem, networks would be designed with full redundancy. Unfortunately, for now at least, cost is a major factor in deciding what services, functions, and equipment need redundancy. The first place to begin planning for redundancy is at the server. The centralized services found on client-server networks make the server an ideal candidate for redundancy, because it is the most cost-effective solution. It is much cheaper to add redundancy to one or a few servers than to an office of workstations.

Multiple hard disks and power supplies are redundant hardware features that are found on file and application servers. Due to the number of moving parts and the sensitive nature of these parts, they have been identified as two points of weakness in a server. The hard disks are particularly susceptible to failure over time. Adding more hard disks is just one part of redundancy in the storage system.

disk mirroring
A disk-redundancy technique that uses a primary disk on which data is stored and a secondary disk of equal capacity to which an exact duplicate of the primary disk data is copied.

disk striping with parity
A hard disk fault-tolerance technique that uses a minimum of three hard disks of the same size. Data is written across all three hard disks, along with some error-correction information called *parity*. If one hard disk fails, a new disk can be rebuilt using the information on the other disks.

As you learned in Chapter 4, adding redundancy at the disk level is called RAID (Redundant Array of Inexpensive Disks). A complete RAID system includes a RAID controller card to manage the hard disks. RAID software on the controller card is used to configure the RAID device and two or more hard disks, depending on the RAID level selected.

RAID level 1 is also referred to as **disk mirroring**. In disk mirroring, all incoming information is written to both disks simultaneously, but data is accessed only from the primary disk.

The advantages of disk mirroring are:

- Only two hard disks are needed.
- An exact replica of the main disk is maintained.
- If the primary disk fails, the second disk is automatically available.

The disadvantages of disk mirroring are:

- Both disks must be of equal size.
- It doesn't offer any performance gains.
- It's expensive because each megabyte of data takes twice as much disk space.

Instead of disk mirroring, RAID level 5 uses **disk striping with parity**. RAID level 5 requires a minimum of three hard disks of equal size. As data is received, the RAID controller writes data across all three drives simultaneously. No one disk has the same information. The parity information is usually an extra byte that is added to every eight bytes of data for error correction. This way, if one of the disks fails, the information on the other disks can be used to re-create the disk that failed.

The advantages of disk striping with parity are:

- Writing to the disks is faster than with disk mirroring.
- A failed disk can be recovered from other disks.
- Parity information is recorded on all disks for error correction and to rebuild the data if a disk fails.
- The cost per megabyte is less than it is for disk mirroring.

The disadvantages of disk striping with parity are:

- It requires special hardware.
- A minimum of three disks of the same size is needed.

Network Management

Performance Monitoring

After the basic components of the network are installed and running, it is time to get into the next phase of network management. Performance monitoring involves several tasks, each of which is intended to be completed with the same goal in mind—to stay informed about the health of the network. This will include evaluating the performance of the network and workstations. Performance monitoring is also a critical procedure for identifying problems on the network. The faster the problems can be identified, the sooner a solution can be decided upon and appropriate action taken.

baselining
Creating a level of reference to use in a performance analysis.

Performance monitoring involves several tasks:

- Setting baselines for network performance
- Analyzing network traffic performance
- Assessing server hardware and software performance
- Documenting performance

Baselining: Setting the Starting Point

If you don't know how the network performs under normal conditions, it will be difficult to determine if it is performing poorly. **Baselining** is a way to set a starting point for evaluating performance. Baselines vary on different networks because of the number of variables involved. Performance can be affected by the network protocol, the speed of the workstations and servers, and the speed of the network. The baseline for a network should be an average of performance measures.

Baselining provides a way to compare data on network performance to an accepted average. If the comparison indicates a significant decrease in the efficiency of the network, the network administrator knows there might be a problem. Without baseline information for this type of analysis, the network administrator won't know when the network is performing poorly until the network comes to a screeching halt.

The first step in performance monitoring is testing the cabling infrastructure. If the network you are administrating uses Category 5 cable, you should have documentation of the Category 5 certification tests. If not, you can use a cable tester to determine the maximum throughput possible on the cable. There are several cable testers that will test the cable and save the test results to a file to print.

Once the testing is complete, you should have an idea of the maximum performance capabilities of the cable infrastructure. With this information in hand, you can set a baseline of network performance by calculating any cable limitations that may exist.

Chapter 13

Analyzing Network Performance

Network communication is a very complicated process. Nowhere is this more evident than when collecting data on network traffic. Analyzing network performance is a two-part process. The first step is to collect data as it is transmitted on the network. You do this by running a network-analysis program that will intercept all information transmitted on the network, whether it is intended for you or not.

The second part of the process involves you, the network administrator. Your responsibility is to evaluate the information for any interesting information. This may include but not be limited to:

- **The types of traffic on the network.** As you will see in the next point, some network-management protocols can create a lot of additional, possibly irrelevant, traffic.

- **The protocols most frequently used.** Some protocols are less efficient than others. AppleTalk, for instance, consumes large amounts of bandwidth, sending out messages regarding devices that are available in the Chooser. On the other hand, a network with a lot of TCP/IP traffic may be a good thing. If your analysis of protocols in use reveals an excessive amount of AppleTalk traffic (or other protocol), you may want to determine the source of the traffic, and determine whether or not the traffic can be controlled or replaced with a more efficient protocol.

- **The frequency of collisions.** Collisions on Ethernet networks are indicative of a network that is saturated with too many devices competing for access.

TIP

As you learned in Chapters 9 and 12, a switch might be a solution to the problem of excessive collisions. Each port on a switch will segment the network into separate collision domains.

- **The percentage of frames with errors.** As frames are received on a device, they are checked for errors. Ethernet frames may have errors because of garbled data or because the frame is fragmented. Packet errors may indicate that a network card is malfunctioning.

- **The devices transmitting the most packets.** Workstations or other devices transmitting an unusually high number of frames in comparison to other computers may deserve your attention.

Network Management

Monitoring the Network

There are several tools available for collecting data on network performance. For the purpose of this demonstration, we'll use AG Group's EtherPeek software, a performance-monitoring program.

In the window shown following, EtherPeek displays details about several packets it has captured off of the Ethernet network. Of special importance are the type of packet and the source of the packet. Alone, the type of packet and who sent it may not be of much interest during the analysis. But when calculating thousands of packets, this information becomes critical. From just these two characteristics, we can determine the protocol that is most heavily used and the source address that is sending out the most packets. The results may be helpful in identifying a problematic protocol or a sending device that is transmitting a large number of packets for no apparent reason.

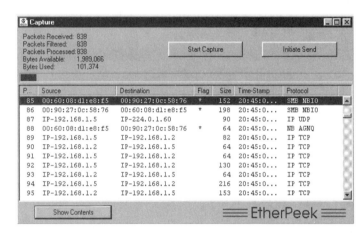

Use of the Protocol Summary report allows you to determine which protocol causes the majority of the traffic on the network.

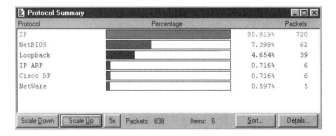

You can save the data collected from your capture of packets to use later to compare to other reports. You also can perform the capture several times during the day to determine peak usage on the network.

Monitoring Server Performance

Network administrators often view servers as the nerve center of the network. When performance lags or a problem occurs, it usually involves the server, and not the network protocol or cables. On the server, there are particular areas that should be monitored regularly. These areas include:

CPU usage   Monitor CPU usage closely. If CPU usage is regularly over 70 percent, it is highly probable that the CPU is unable to process all the requests that it is receiving.

Swap-file errors   Servers use part of the drive space for temporary storage of data that typically would be saved in memory. If the server does not have the correct location of the data it is looking for in the **swap file**, it causes a **fault**.

Network traffic to the server   The demand on the server can be measured by the amount of traffic it is sending and receiving. A high volume of traffic can impact CPU usage as well.

Disk read/write   Excessive reads and writes to the hard disk may be due to the server constantly using the swap file for memory, indicating that more RAM is needed.

There are many products available on the market for monitoring server performance. The following exercise uses the Performance Monitor program that is included with Windows NT Server 4.

Documenting Performance

In the last section, you completed an exercise that taught you how to monitor server performance. In Chapter 12, we stressed the importance of documentation in the LAN design process. Keeping documentation is still important as you manage your network. Documentation can include reports that are created from applications such as Performance Monitor.

Other applications generate log files automatically. Some systems, such as UNIX, require that you run a system log server to collect system-wide information on what events are occurring on the server. These log files can include the name, date, and time of logins. They can record what files were accessed successfully; what attempts for accessing files failed; and what files were changed, created, or read. The system log can also be very important for security reasons. Details about who has logged in or who is attempting to log in to your system can be identified. Hopefully, the information in the systems log can identify any security problems.

---

swap file
A part of the hard disk that is used as a temporary storage space when the computer has used all available RAM.

fault
A memory error that occurs when the CPU is unable to find the location of data stored in the swap file.

Network Management

Test It Out: Configuring Performance Monitor

1. From the Start menu, select Programs ➢ Administrative Tools, and click on Performance Monitor. A blank Chart window will appear.
2. Click the Edit menu and select AddCounters.

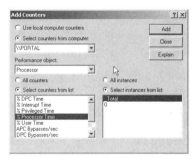

3. In the Add to Chart window, click on the Object pull-down menu and select Processor. In the Counter field, select % Processor Time. Click Add. Repeat step 3 using the following table. When you are finished, click Done to view the graph that is being created in Performance Monitor.

Object	Counter
Memory	Page Faults
Network Interface	Bytes Total/sec
Server	Bytes Total/sec

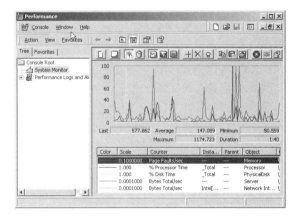

Viewing performance information in Performance Monitor requires setting an object and a corresponding counter. An object is a component or function that has the capability of being monitored. This capability comes from counters. A counter is just a way of measuring the performance of an object.

Documentation is indisputably the weakest link in a network administrator's arsenal. Because a large part of your job as a network administrator is to support users, much of your valuable time is spent troubleshooting and catering to the needs of those users. But you need to document events, especially problems, as they unfold (or at least at the end of the day). If you wait until the end of the week, you are bound to forget most of what you learned and solved.

You don't need to spend *all* of your time documenting the health of the network. Try these techniques to minimize your responsibilities while maintaining a current, detailed, and accurate set of documentation:

- Keep a journal of the day-to-day activities on the network in a composition book.
- Create a file system for organizing and maintaining documentation.
- Provide users with a binder to log problems they encounter.
- Create a web-based support database for users to record their problems.
- Develop a checklist of documentation tasks that need to be performed weekly, monthly, and quarterly.

Network Management Systems

A **network management system (NMS)** uses a combination of hardware and software to monitor and manage devices on the network. Network administrators rely on the network management system for up-to-date information on the health of the network. Whether it's performance, inventory, configuration changes, or notification of network failures, a complete NMS can reduce the time involved in managing the network. Unlike a monitoring program that runs on a server, NMS software will monitor network devices from one workstation, regardless of the vendor of the network device.

In a network management model, there are objects that represent elements of a network device, such as a port on a hub or system information. These objects and the corresponding attributes that provide details about them can all be monitored and changed remotely. Hubs that have been designed to participate in an NMS have several objects. An object such as an Ethernet interface can be asked to provide detailed information about its configuration and performance.

In order for the network management system to gather information about a device, it has to know about the device's attributes. Attributes can include make, model, software versions, performance information, configuration settings, and the technical support contact information. Many of the performance attributes can be recorded temporarily by the device for use by the NMS. If the value for the attribute exceeds a certain percentage, or **threshold**, the network management system can alert the network administrator of a possible problem.

The Management Model for TCP/IP Networks

TCP/IP networks have been around for a long time, and recently have grown into the complex network we know as the Internet. Managing TCP/IP networks was once an easier task. Network administrators could use the Ping program to test whether a network device was available on the network. As the TCP/IP networks have grown into complex networks, utilities such as Ping have become inadequate for managing the network or simply can't be used due to security restrictions enforced by firewalls.

In 1990, RFC 1157 defined the Simple Network Management Protocol (SNMP). The protocol was created to help network administrators manage their growing networks. More importantly, network administrators could use SNMP to manage all kinds of network devices from many different vendors. As long as the vendor created software that was compliant with SNMP, the network administrator would have a level of control over the device that was previously unavailable.

network management system (NMS)
The use of hardware and software by a network administrator to monitor and manage the network.

threshold
A value set by the network administrator that, when exceeded, indicates there is a critical situation.

The SNMP management model defined in RFC 1157 is specific to TCP/IP networks; SNMP is included in the TCP/IP protocol suite. When SNMP is implemented on a network, it becomes a very powerful tool for the network administrator. Specific details about hardware and configurations and IP protocol information can be recorded. In the event of a failure or performance problem, the information can even be used to alert the network administrator.

management information base (MIB)
A database that contains the definitions and values of objects for a managed device.

The SNMP protocol is only one component of the SNMP management model. The model includes:

- Manageable devices
- A network management console
- Management Information Base (MIB) agents
- Simple Network Management Protocol

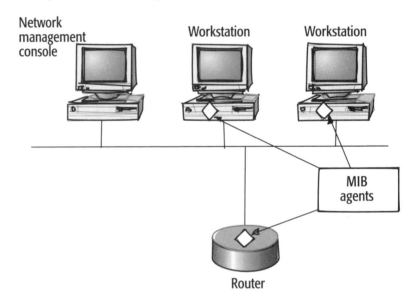

SNMP in Action

The SNMP protocol gathers management information from the manageable network devices. Each manageable device—whether a switch, router, or computer—has information on certain variables that are stored in a database. The variables, called *objects*, include history, hardware, configuration, and status. The objects are all defined in RFC 1155 (on management information). All of the management information objects that exist on the network are collectively known as the **management information base (MIB)**. Each SNMP-compatible device created by a vendor must have a corresponding MIB file that details how to access the object information for that device. Without the MIB file, the management console and the SNMP software will not be able to recognize the device's objects.

Network Management

Object information that is stored in the database—that is, the device status and other variables—must then be collected and transferred to the management console when requested. When a request for information is made or a change is sent to a device, the **agent** software running on the device uses the MIB to collect object information from the database. The agent can then either respond to a request or update the database with information that is received from the management console.

SNMP communicates management information between the network management console and the manageable devices. With SNMP, this is done using UDP, a Transport layer protocol, over IP, rather than TCP.

agent
Software that collects MIB information about a managed device, and provides the information to the management console.

NOTE
Because UDP does not require any confirmation of delivery between the sending and receiving device, an SNMP message is transmitted by the sending device with the assumption that it will be received by the destination device. UDP helps minimize network utilization.

There are many network devices that support SNMP. Due to the demand for centralized network management, most companies include SNMP functionality on their equipment. Hubs, routers, switches, workstations, and servers that are manageable can respond to SNMP commands.

Once you have identified the devices on your network that are manageable, you need to dedicate a computer to act as your network management console. As you will learn a little later, the management console will require some special hardware configurations. In addition, you will need to purchase or download SNMP management software.

Unlike other network-management protocols, SNMP is relatively simple and reliable because it uses only three basic commands. An administrator uses the `get` command to request information from the agent on a manageable device. When the network administrator wants to change a configuration setting on a device, they use the `set` command. The `set` command changes the value of a configuration setting, rather than initiating the command on the device.

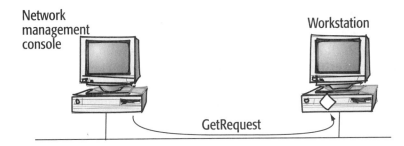

Network management console — Workstation — GetRequest

301

Chapter 13

Here's how an administrator would use the `get` command to get information about a device:

1. From the network management console, the administrator sends a GetRequest for system information from a device.
2. The device receives the IP packet and passes the SNMP data to the agent.
3. The SNMP agent collects the system information and puts it in the right format.
4. The SNMP agent sends back a GetResponse with the system information.
5. The management console receives the GetResponse with the device's system information enclosed and updates the device information in the management console software.

The last SNMP command, `trap`, is used to set a manageable device to automatically notify the management console of a system failure or performance problem. It is common for an attention-getting alarm or alert to be triggered on the network management console when information from a trap is received. However, because SNMP uses an unreliable protocol, it is possible that the management console may not receive a trap. As a precautionary measure, **polling** was added to the function of the management console. At regular intervals, the console will poll manageable devices for problems.

Remotely Monitoring a Network

Although SNMP is widely implemented, it has its problems. Network administrators who rely heavily on SNMP for critical system information run the risk of burdening the network with SNMP traffic. On large enterprise networks that have thousands of manageable devices, network performance can be seriously impacted by the various SNMP commands transmitted between the network management console and the manageable devices as part of the polling process. The problem is further compounded on enterprise networks when the SNMP traffic has to travel over WAN connections.

There is a way to cut down on network congestion caused by polling. RMON is a remote monitoring function that is supported by some devices. RMON is defined as part of the management information base and it has its own objects. Each of the objects detail summary information about the network and devices on the network. Network devices capable of running the RMON functions are called RMON probes. RMON probes do not communicate with the managed devices directly. Instead, the RMON probe collects data from the network in **promiscuous mode**. As data is transmitted on the network, the RMON probe collects the information on what is being transmitted and by whom. The management information

---

polling
A process used by the SNMP protocol to periodically check for updated management information from the devices.

promiscuous mode
The collection of data from the network without other devices having any knowledge of the activity.

Network Management

collected by the RMON probe can then be sent to the network management console. Traffic is greatly reduced because the RMON information from that network segment of the internetwork is communicated directly between the RMON device and the network management console.

Common Management Information Protocol

The Common Management Information Protocol (CMIP) was developed by the ISO in response to the limitations of SNMP. CMIP provides more detailed information about a device, and is in general a more complex protocol than SNMP. Because it is a fairly new protocol, CMIP is not as widely available as SNMP. Cisco is one vendor that does support SNMP and CMIP in most of its products.

Implementing Network Management Systems

Implementing a network management system (NMS) on a network requires thorough planning. Like anything else in networking, it is important to document your plan and everything that you do while implementing SNMP.

As part of implementing an NMS, the management console station should be selected by following certain criteria. First, you need to determine the management software that you will use. If you will primarily be supporting Cisco products, you may decide to use CiscoWorks management software. Otherwise, there are many other products available, including Castle Rock's SNMPc, Hewlett-Packard's OpenView, and Computer Associates' NetworkIT, among others. These programs offer at a minimum these basic functions:

- They allow discovery of nodes.
- They offer support for both IP and IPX.
- They offer a graphical view of managed devices.
- They allow you to generate reports.
- They send alerts in the form of an audible alarm, an e-mail message, or contact via a pager.

When determining the hardware requirements for your management console, follow the minimum system requirements set by the software manufacturer for the management application you have chosen. It is safe to say that you will need a Pentium-class PC with at least 250MB of free hard-drive space and 64MB of RAM. If you are using a network management application that creates 2D and 3D renderings of the network, the hardware requirements will be much higher.

In addition to deciding on hardware and software for your management console, you must also determine what kind of security will be used on the console. The management console is capable of collecting configuration information from network devices, including login names and passwords. Also, the network management console gives whoever uses it the ability to view, change, and delete configuration settings on devices. The management console should be in a secure location such as the **network operations center (NOC)**.

network operations center (NOC)
The central location in a network where all management is maintained, and core devices such as routers are located.

Network Management

Troubleshooting

Troubleshooting is a process of identifying that a problem exists and then determining a solution to remedy the problem. Troubleshooting computers and networks is a large part of a network administrator's responsibilities. Being a good troubleshooter takes more than just solving the problem. The network administrator needs to:

- Approach troubleshooting as a process
- Document problems and their solutions
- Be efficient in responding to and diagnosing problems
- Develop a resource kit of information and tools

When a problem exists, whether it is on a workstation or on the network, you, the network administrator, need to be attentive to the problem. Users won't expect a miracle, but the longer it takes for you to respond, the longer it will be before the users can get back to work.

Being an efficient troubleshooter involves adopting a process that systematically narrows down the possible sources of problems to just a few, or even one. Computers and computer networks are extremely complex, with hundreds or even thousands of variables that could be causing an error or malfunction. In heterogeneous networks, in which there are several different makes and models of hardware and software, the process of elimination becomes critical because the number of potential causes of a problem is even greater.

The elimination process is like a funnel. When you first begin eliminating possibilities, you attempt to eliminate a large group of factors immediately. For example, when initially approaching a problem with a computer, you may be able to determine that the problem exists solely on that computer. Therefore, you can eliminate all other computers, the network, and the servers as possible causes of the problem. The next step in the elimination process involves fewer possible sources of the problem, and you can thus narrow the focus of the search until eventually there are only a few possibilities left. Using an elimination process is not the same as just keeping a checklist of things you have tried. It is more efficient to use a checklist in conjunction with the elimination process that has been described.

Documentation for Troubleshooting

If you have ever done any troubleshooting, you know that one of the keys to solving a problem efficiently is having the information you need close at hand. The documentation you'll use when troubleshooting comes from several sources;

some of it is created on-site, and some of it is provided online or in other formats from vendors. As a network administrator, you will need the following items as part of your resource kit:

Journal   This component of your network administrator's resource kit is one of the most important pieces of documentation. Not only for planning, the journal should be used to document problems and the solutions that can then be used in the future.

Problem log   A notebook, binder, or electronic system should be available for users to document problems.

Books   Create a library of computer, software, and networking books that you keep close at hand for researching problems.

Vendor documentation   Most vendors offer support documentation, either online or on compact disc. Take the time to bookmark technical support websites in your browser. Nearly all solutions to vendor-specific problems can be found on the vendor's website.

Software updates   Know how to find software updates from vendors. Software updates are created almost daily as bugs are found and incompatibilities between different pieces of software are discovered.

End users can be the eyes and ears of the network administrator—if they are given the right tools. Depending on the size of your organization and your resources, consider the following options:

Support database   A support database can be used to collect the specific details of a problem that a user experiences. Like a problem log in a notebook, the database offers far more flexibility to search, generate reports, and even allow you to respond to questions.

E-mail   Assuming that the problem isn't that the e-mail server crashed, users can send questions and details of problems via e-mail. A savvy network administrator can organize the e-mails for future use.

Without user documentation, troubleshooting can be very difficult. Users will not only need a system to report their problems, but they will also need training on how to document a problem well.

A Layered Approach to Network Troubleshooting

Network troubleshooting can be simplified by using the OSI model to pinpoint problems. As you recall, the OSI model is a seven-layer model that was created to allow developers to build hardware or software that operates at a given layer

Network Management

and is interoperable with the layers above and below it. The OSI model also makes troubleshooting easier. As each layer is tested, the problem can be narrowed down until eventually the problem is identified at a given layer and a specific solution can be presented.

Troubleshooting according to the OSI model can begin at the top layers (4 through 7), the middle layers (2 and 3) or the bottom layer (1). Once again, where you begin depends on the information that is given to you regarding the problem.

Network Application Errors

Application errors can occur on the workstation or server. These errors cannot occur on any other type of network device, unless the device is capable of running an application service. Application services include FTP, HTTP, DNS, DHCP, and SMTP, among others. When errors occur at layers 5–7, they usually manifest themselves in the form of errors on the screen or a system that won't function. It some cases, the application won't respond. For example, Internet Information Server, a Microsoft web server for Windows NT Server, may not be responding to end-user requests for web pages. To confirm this, you can try opening a web page from a workstation. If you don't have any luck, you can open a web browser on the server and try accessing a website. If the site doesn't respond, you can temporarily stop the service and then restart it.

Application	7
Presentation	6
Session	5
Transport	4
Network	3
Data link	2
Physical	1

If an end user receives an error message, it is important to document the message or any error codes that appear on the screen. Also, you should ask the user to determine what they were doing when the problem occurred.

Other tips for troubleshooting application problems:

- Applications that unexpectedly quit might need to be reinstalled.
- If you can access the Web, but you are unable to access a new website you registered for your company, it is probably because the DNS database was not updated.

- Regularly check for software updates from the vendor. The problems that you are experiencing may have been fixed in an updated version.
- Conflicts between applications usually occur after a new application is installed. First, uninstall the application; then check with the manufacturer for known incompatibilities.

Protocol Problems

Troubleshooting protocol problems varies in complexity, depending on the protocols in use at layers 2, 3, or 4. Layer 2 protocols include Ethernet, Token Ring, ARCNET, FDDI, ATM, and WAN layer 2 protocols. Layer 3 and 4 protocols include TCP/IP, AppleTalk, IPX/SPX, DECnet, and NetBEUI.

Application	7
Presentation	6
Session	5
Transport	4
Network	3
Data link	2
Physical	1

Networks that use the TCP/IP protocol suite are susceptible to more problems than AppleTalk and IPX/SPX networks. One of the top reasons for this is that AppleTalk and IPX networks automatically assign network addresses to workstations. TCP/IP networks can easily be incorrectly configured because of the vast set of numbers to keep track of. Manual errors can be the cause of conflict IP addresses.

Troubleshooting Protocol Problems

At the first signs of a conflict IP address, you can usually determine that it was caused by one of two things:

1. If some or all workstations have been manually assigned IP addresses, then the IP addresses were allocated incorrectly. You can identify the errant workstation using the MAC address that appears in the error message.
2. If you are using DHCP to dynamically assign IP addresses, there may be a problem with the DHCP service. This usually happens in situations where the DHCP server allocates some IP addresses and then has to be restarted for some reason. When the DHCP service restarts, it begins

Network Management

allocating the same IP addresses that were already assigned to workstations. Try restarting the workstations. You can also run Winipcfg (95/98) or Ipconfig (NT), release the IP address, and then reboot; this usually takes care of the problem, too.

Besides conflict IP addresses, other problems with TCP/IP can occur on your network. A common problem is that servers are not available or cannot be reached. As you learned in Chapter 4, two utilities, Ping and Traceroute, can be used to identify whether the device is not responding or if there is a failure in the path that the information takes to reach the destination.

Ping   This utility tests connectivity between two devices. Ping will also provide valuable information about the amount of time it took to receive a reply and the number of successful tries. You can also use Ping to test the address of the computer that you are working on; either ping the IP address assigned to the computer or use the loopback IP address 127.0.0.1.

Traceroute   Traceroute can be used to determine the path that a packet takes to reach its destination. You can map the path by following the IP addresses that are returned in the trace. If the trace stops before it reaches the destination, you can then pinpoint the problem at that device.

Besides the TCP/IP protocol, you may have to troubleshoot problems with Ethernet or another layer 2 protocol. Outside of problems with the network card, identifying layer 2 problems can be complicated. On Ethernet networks, problems arise from collisions and broadcasts. If the network is suffering from problems such as collisions, errors, or excessive traffic, a network-monitoring program can be used to determine their severity.

Problems with Network Media

At layer 1, the issues revolve around physical problems. These include cable, electrical power, devices such as hubs and NICs, and all other points of termination.

Application	7
Presentation	6
Session	5
Transport	4
Network	3
Data link	2
Physical	1

Network cable problems are usually resolved when the network is installed, at which time the network media should be tested and documentation provided on each of the tests. Because the cable runs that go from the jacks in the wall to the wiring closet are permanent, there is usually no reason for the cable to not work. To make sure that the cable was installed correctly, you can use a cable tester. A basic cable tester will test the wires for continuity. If the cable is able to transmit a signal, the wire is considered adequate for data transmission. More-sophisticated testers will identify problems between pairs such as:

Short   Two wires are touching, creating a short in the circuit.

Reverse   One pair is flip-flopped with another pair.

Cross   The two wires in a pair are flip-flopped at one end.

Split   One of the wires in a pair is in the wrong location.

Problems can occur in the patch cables used to connect a workstation to a wall jack. Patch cables are often mistreated and can become damaged. If the workstation that you are troubleshooting is unable to see the network at all, and you have checked that the software is configured correctly, you can quickly test the patch cables to make sure that they are in good working order. The problem may be as simple as a disconnected cable.

A less-common layer 1 problem is that hubs and NICs can fail over time. If the cable passes its tests, and the software on the workstation is functioning, the NIC or a port on the hub may have failed.

Review Questions

1. What are three ways to provide fault tolerance on a network?

   _____

2. Why are Level 1 support personnel best-suited to solve desktop support problems?

   _____

3. How do remote desktop control applications reduce troubleshooting time for support staff?

   _____

4. What are the advantages of using a Travan drive over a DAT drive?

   _____

5. What is the difference between a differential backup and an incremental backup?

   _____

6. What are two ways to monitor server performance?

   _____

7. What information can be gathered using a network-analysis program?

   _____

8. What is the purpose of the MIB file?

   _____

9. What items should be included in a network administrator's resource kit?

   _____

Terms to Know
- agent
- archive bit
- autoloader
- baselining
- bottleneck
- Digital Linear Tape (DLT)
- disk mirroring
- disk striping with parity
- downtime
- FAQ (frequently asked questions)
- fault
- fault tolerance
- latency
- management information base (MIB)
- MIC
- network management system (NMS)
- network operations center (NOC)
- polling
- promiscuous mode
- redundancy
- SCSI
- SuperDLT
- swap file
- threshold
- Ultrium (LTO)

Chapter 14

WANs and Internet Access

By now, you should be tinkering with your network, trying out what you've learned, and hopefully asking lots of questions. In order to be well-prepared for your CCNA studies—and for your responsibilities as a network administrator—you also need to understand wide area networks.

Most people aren't familiar with wide area networks, but whether they know it or not, they use a wide area network to access the Internet. A network administrator, however, must understand the technologies that make wide area networks possible.

In this chapter, wide area networks and Internet access will be demystified as you learn about the following topics:

 What wide area networking is

 How Internet access works

 What telecommunications services are available

 How to determine Internet access needs

 Steps in implementing a WAN connection

Chapter 14

Wide Area Networks

In Chapter 1, you were introduced to wide area networks (WANs). A WAN is a network that is created using either the services of the telephone company or fiber-optic cable owned by large telecommunications companies. The physical line leased from the telecommunications carrier is used to connect LANs separated by large geographical distances. In order to make the connections, WANs require the use of special WAN protocols and devices. Although many network administrators only have WAN connections for the purpose of Internet access it is common for larger companies to have WAN lines to connect offices separated by a few to thousands of miles. And with the cost reduction associated with creating WANs it is now feasible for smaller companies, like a wine maker for example, to connect the business office LAN to the winery LAN.

Internet access, a.k.a. wide area networking, is just another way of connecting resources to the LAN. Although Internet access has become increasingly important, if not necessary, in many situations, you don't have to have Internet access to have a LAN. Wide area networking introduces a whole new level of complexity to internetworking. Different technologies, from protocols to devices, are needed to make WANs possible. Another significant issue is that you must coordinate and manage your network together with outside organizations—specifically, your local telephone company or telecommunications provider and an Internet service provider. As you will see later, there are potholes in the road to the Internet.

The characteristics of a WAN are:

- Inexpensive WAN connections are low bandwidth, with speeds typically between 56Kbps and 1.5Mbps. Some companies, universities, and the government can afford high-speed WAN connections, 45Mbps to as much as 1Gbps, but at a significantly greater expense.
- A single WAN connection is shared by all devices on a LAN unless a second link is installed for redundancy.
- WANs are used to cover large geographical areas.
- WAN technologies include Point-to-Point Protocol (PPP), Ethernet, Asynchronous Transfer Mode (ATM), Integrated Services Digital Network (ISDN), Digital Subscriber Line (DSL), **dial-up** modem access, and Frame Relay.
- WAN devices include routers, modems, and WAN switches.

WANs have come a long way in just the past 10 years, but their fundamental structure has not changed much. Extending a network using a WAN connection still requires that you lease a telecommunications circuit. Whether the physical circuit comes from the telephone company as a copper wire; or from another

---

dial-up
The process of connecting to the Internet using a modem and telephone line.

WANs and Internet Access

company using wireless, cable communications, or fiber-optics, the WAN connection involves an outside party.

As a network administrator, you need to know how the telecommunications part of the process works. What would appear to be a simple example of dial-up Internet access turns out to be a complex maze of equipment and technologies.

This chapter will focus on WAN technologies and options for small- to medium-sized business. This will keep you focused on the goals of the CCNA certification.

Chapter 14

Connecting to the Internet

In all likelihood, your first exposure to planning and implementing a WAN will be deciding on the type of Internet access for your company. You probably already have some experience selecting a WAN connection if you have Internet access in your home. It doesn't matter if you have been using the Internet at home for the past 10 years or if you have just ordered DSL for Internet, you are dealing with a WAN connection to the Internet.

When you connect to the Internet using DSL, there are several processes that must take place in order for the connection to go through. First, you may be required by your ISP to use special software to log in. Although a DSL user who has to use software still has an "always on" connection, they are unable to make use of their high-speed access without verifying their identity using a login name and a password. In any case, the DSL modem or router must negotiate a connection with the DSL router at the CO. (You will recall that the phone line from your house to the phone company goes to the central office in your neighborhood first.) Unlike a dial-up modem, the connection between the devices occurs as soon as the device in your home is turned on. You don't tie up a phone line and there is no loud crackling noise as with dial-up modems.

Like a dial-up modem, DSL modems use a regular two-wire phone line called the **local loop**. The two-wire, twisted-pair cable extends from the residential **premise** to the telephone company's central office exchange, also called the **local exchange carrier (LEC)**.

At the CO, the loop ends at a switch that contains many interfaces for all of the phone lines. The switch in the CO uses **switched circuits** to connect two voice conversations. (If you have DSL, it is usually a clear indication that the phone company has replaced the older voice-only switches with digital switches that support both voice and data.) If the call is destined for a **subscriber** at a different CO, it is forwarded to the **metropolitan exchange**. The call is eventually forwarded to the CO nearest the ISP.

local loop
The two-wire copper telephone cable that runs from a residential or commercial location to the central office of the telephone company.

premise
The location of equipment and wiring at the end-user site.

local exchange carrier (LEC)
The local telephone company and the CO switches that provide local service.

switched circuits
Once a mechanical feature, switched circuits are used to create logical connections between two points on a switch in a CO.

subscriber
A home or office that utilizes the public telephone network.

metropolitan exchange
The main telecommunication switches found in a city to centrally connect all COs.

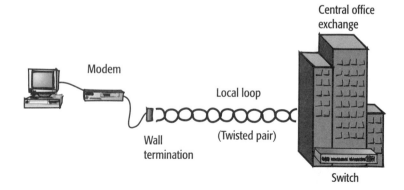

WANs and Internet Access

You'll remember from Chapter 7 that frequencies and modulation are characteristics of a signal. A technology like DSL exploits the use of frequencies to allow a DSL line to coexist on the same line with a phone. Because the two signals, one voice and the other digital, are on different frequencies they don't interfere with one another.

Connecting to Your ISP

Between the local exchange carrier and your ISP lies a four-wire cable used to handle all of the traffic of DSL and dial-up customers.. This is called a **multichannel trunk** which is often a T1 line or better. In the case of dial-up users, the phone calls destined for the ISP are **multiplexed** on the trunk. When a call reaches the ISP, it is **demultiplexed** and split into an individual line that connects to a modem.

multichannel trunk
A four-wire cable, used for dedicated leased lines, which is capable of supporting up to 24 voice conversations or 1.536Mbps of data.

multiplex
Sending multiple voice or data signals across a single wire by inserting bits of information from each of the signals into a frame.

demultiplex
Separating the bits of information from a multiplexed frame back into the individual voice or data signals.

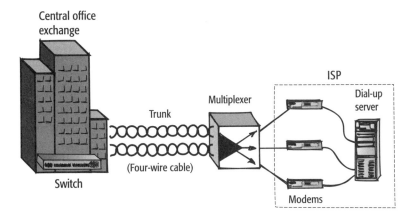

NOTE

Although many ISPs may use a four-wire cable, called a lease line, to connect to the local exchange carrier, most of the largest companies utilize fiber-optic cable. LECs, like Pacific Bell in California, have upgraded their infrastructure to fiber-optic cable in most urban areas in an effort to support the growing demand for both data and voice traffic.

The connections created by DSL users don't look that much different than dial-up users. DSL users still make their connections to the CO where they are connected through high-speed trunks to the ISP. Unlike dial-up users, the connections are always digital and the signal is never multiplexed or demultiplexed. The DSL traffic goes straight from the DSL modem through the CO to the appropriate ISP.

Planning Internet Access

Planning Internet access for a business or other organization is more complicated than deciding on an Internet connection for the home. When choosing an ISP for home Internet access, the driving force behind the choice is usually cost followed by convenience and finally performance. But for a business, the cost for the service can often be outweighed by the needs of the company. Businesses have to balance the need for bandwidth, the reliability of the technology used to connect, and the quality of service offered by the Internet service provider. Upon first accepting the mission to plan Internet access for your company, you must answer some basic questions:

- How will the connection be used?
- How important is the connection to the company?
- What infrastructure already exists?

First, you will need to consider future applications of the WAN link—how the connection will be used. A few applications that may influence your decision are listed as follows; each has been selected based on the growing demand for these services and their impact on performance:

Video conferencing/streaming Most types of video require a constant flow of data, rather than the bursts that are typical on the Internet. If you need video now or in the future, make sure you are investing in a technology that will support it.

E-commerce If your company has a successful **e-commerce** site or is planning to start one onsite, it may seriously impact the performance of your WAN connection.

Virtual private networks (VPNs) Your company may want to implement a VPN to give employees external access to the company network via the Internet. This could create a significant demand on your WAN.

In addition to determining how the connection is to be used, you will need to identify the importance of the connection to the company's success. Does the company need the connection up and running no matter what? Does the company lose money if the connection goes down? If you answered "yes" to one or both of these questions, you will want to select an Internet connection that is reliable and use a technology that has been well-tested.

One final task before you search out an ISP is to determine what equipment you have. If you do have existing equipment, such as router or digital modem, you need to determine whether it will be of use in the new system.

---

e-commerce
Short for *electronic commerce*. The buying and selling of goods or services over the Internet.

Telecommunications Services

To plan an Internet connection for your company, you need to be familiar with the telecommunications services available. Besides knowing about the PSTN (public switched telephone network), which was covered in earlier chapters, you'll need to understand the term **T-carrier**, an important concept in WAN communications.

T-carriers were originally created by AT&T as a way of transmitting multiple voice conversations over a single four-wire line called a *trunk*. A T1 is a dedicated **leased line** that can transmit at a rate of 1.544Mbps. Other types of T-carrier lines include T2, which transmits at 6.132Mbps, and T3, which transmits at 44.736Mbps.

Circuit Switching

Circuit switching is used to create channels for voice conversations on an as-needed basis. When a phone call is made, a circuit switch at the CO creates a logical connection between two points for the duration of the call. Anyone who wants to make a phone call must connect to a circuit switch. Because only a limited number of lines are available, it is possible for all lines to be busy. This usually isn't the case with T1 lines because the circuit switch is maintained indefinitely between the company and the Internet service provider.

For voice transmissions, every phone conversation that is created requires a 64Kbps channel. Of the 64Kbps, 8Kbps are used for framing, leaving a total of 56Kbps.

Circuit switching is of particular importance because the technology supports analog and digital signals. A single digital channel is called a **DS0**. A multichannel trunk has a total capacity of 24 DS0s.

If an entire trunk is in use for voice, data, or both, the line is called a T1. The "1" means that all 24 DS0s are in use. When ordering a T1 line for Internet access, you should request a clear-channel T1. A clear-channel T1 means that all DS0s are available for use all the time for data transmission. The result is a total T1 capacity of 1,536Kbps with only another 8Kbps used for framing.

Types of circuit-switched services include:

- Switched 56
- Switched T1
- ISDN

T-carrier
A technology developed by AT&T that was designed to provide multiple voice conversations over a single four-wire cable called the *trunk*.

leased line
A four-wire copper cable that is leased from the telephone company. The digital line provides high-speed digital transmission between two locations.

DS0
A digital channel used for data or voice transmissions.

NOTE

You can use a single trunk for voice and data by allocating a set of DS0s for voice and leaving the rest for data. Some equipment will allow you to allocate channels dynamically.

Chapter 14

Packet Switching

Packet switching is a technology that is used to transfer data between two points over a shared medium. Unlike circuit switching, which dedicates a connection between two points, packet switching sends packets of data across the network using the most direct path. Due to the burst-like nature of data transmission, packet switching is an efficient solution.

cell switching
A form of switching, similar to packet switching, that uses fixed 53-byte cells to transport data.

In circuit-switching connections such as T1, the cost is assessed for the total capacity of the line and not the bandwidth used. Packet switching is usually billed based on usage.

Some types of packet-switching networks include:

- ISDN
- Frame Relay
- X.25

Due to the rapid fall in the cost of circuit-switched T1s, the cost savings for packet switching are not quite as broad as they once were. In addition, the introduction of new technologies, in particular **cell switching**, has lowered the cost of high-speed Internet access even further.

NOTE

ISDN supports bearer services that include packet switching and circuit switching. (Bearer services are types of telecommunication services that are concerned with moving information from one location to another.)

WAN Technologies

Now it is time to take a look at the WAN connection technologies. To implement any of these WAN technologies, there must be coordination among your company, the telecommunications company, and the Internet service provider. Otherwise, you may find that the ISP offers a service at an excellent price, but that the telecommunications company doesn't provide service for the technology in your area.

NOTE

With the deregulation of local services, you may be able to evaluate prices for dedicated leased lines and other telecommunications services from a number of companies, including your local Bell company. The main difference between selecting a competing local provider or the local phone company is that the competing local provider is still leasing from the local phone company, just at a lower rate.

WANs and Internet Access

Point-to-Point Protocol (PPP)

A point-to-point connection consists of a dedicated logical connection between two points over a leased line. PPP supports nearly all network protocol types and is most frequently found in dial-up connections.

The advantages of PPP are:

- It's well tested and implemented.
- It offers excellent throughput.
- It supports most network protocols.
- It's simple to implement.

The disadvantages of PPP are:

- It's less flexible for expanding to multiple sites.
- It's expensive ($300–$1500 per month, depending on bandwidth).
- It requires dedicated leased lines.

Digital Subscriber Line (DSL)

DSL may be the best alternative for Internet access for small- and medium-sized businesses as well as residential locations. DSL is a broadband technology based on ATM (Asynchronous Transfer Mode). DSL runs on existing telephone lines, but the service requires an infrastructure upgrade at the CO.

DSL comes in two flavors: Asymmetrical DSL (ADSL) service offers a higher download transfer rate than upload rate—for example, 384Kbps/144Kbps. Symmetrical DSL (SDSL) transfers at the same rate in both directions.

The advantages of DSL are:

- It's very affordable ($50 and up, depending on service).
- Currently, it supports speeds of up to 6Mbps for downloading and 1.5Mbps for uploading.
- It runs over normal telephone lines.
- Data and phone calls can exist on the same line simultaneously.

The disadvantages of DSL are:

- Availability and quality of service depends on the condition of the line and the distance from the office building to the CO. This distance is usually less than 1500 feet.
- DSL service is not available everywhere.

Chapter 14

Configuring a DSL Router

Although a few manufacturers offer an integrated DSL router with DSL modem, most small business and home users will probably already have a DSL modem from their DSL provider. If this is the case, they can use a DSL router to provide some low-level security. Some newer models offer wireless connectivity, as is the case with the Linksys DSL/Cable wireless router used in this exercise. Similar products do exist for SMC, NetGear, and Cisco, among others.

1. Follow the instructions from the manufacturer for installing the DSL router. This will involve connecting the DSL modem to the router using a network cable and plugging in the power.

2. Connect your computer to the DSL router using a network cable.

3. Configure the TCP/IP settings for your computer to obtain an IP address dynamically.

4. Open a web browser, put the following IP address in the Address bar: 192.168.1.1, and press Enter on the keyboard. This is the factory default IP address used by the router.

5. When the login prompt appears, enter the username and password given in the documentation provided by the manufacturer.

6. In the following Quick Setup window, you can change the IP address of the LAN port from the default 192.168.1.1, and enter the WAN IP address and subnet mask provided by your DSL or cable Internet provider.

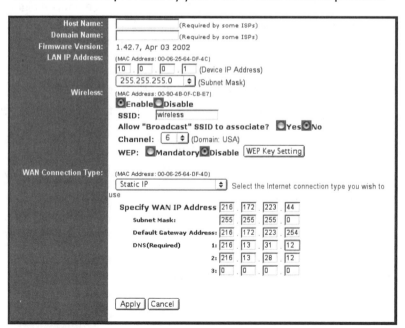

WANs and Internet Access

7. Click on the DHCP tab to configure DHCP settings for computers that will be accessing the Internet from your network.

8. Enter the starting IP address (usually 1) and the number of users. Any number greater than the total number of network devices should be plenty.

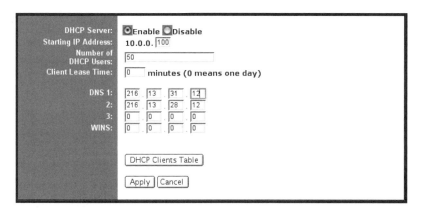

9. Enter the DNS servers that will be used. If you don't think you have DNS servers on your network, then use the ones provided by your Internet provider.

NOTE

In this example, I could have set the Number of DHCP Users as high as 154. Because my starting number is 10.0.0.100, the last possible number available for use on the network is 10.0.0.254. If I had started with 10.0.0.2 then there would have been 253. The 10.0.0.1 address is in use by the LAN interface of the router.

Data Over Cable

Although many shied away from the use of Cable for Internet access, it has emerged as a viable broadband option for home Internet users. The data over cable technology is based on the ITU-T J.112 standard, Data Over Cable System Interface Specification (DOCSIS). Data over cable is better known as cable Internet technology. Over the past several years, cable companies have updated their television cable infrastructure to provide two-way broadband communications. Cable modems allow users to use the existing television cable coming from the street for high-bandwidth data transmission in addition to television. The great advantage of cable modems is the bandwidth that you get for the price.

Chapter 14

NOTE

Cable technology is usually limited to residential buildings. If you have a home office or telecommute, cable will definitely be the lowest-cost solution. If you are a business, you will first need to have a TV cable connection. Check with your local cable company to see if they offer the service at your site.

Internet access over cable lines can run at a whopping 10Mbps. If cable access is an option in your area, it is worth investigating data over cable; there really is no comparison for the price.

The advantages of data over cable are:

- It can provide high-bandwidth Internet access at 10Mbps.
- You can maintain your cable television over the same cable.
- It's extremely inexpensive (less than $50 per month for a home connection).
- Future uses will probably include telephone access on the same cable.

The disadvantages of data over cable are:

- It's not readily available in all areas.
- The technology is a shared medium, meaning that if there's a lot of traffic on your cable segment, you may be receiving at speeds much less than 10Mbps.

DSL versus Cable Internet

The telephone companies have quickly adopted DSL as an affordable high-bandwidth solution for residential users. Concurrently, cable companies—the largest of which are now owned by AT&T—have already upgraded large portions of their infrastructures to support high-speed data, local and long distance telephone service, and cable television all on one cable line. Both options work well and are proven. The decision of which is better may ultimately come down to customer service, not technology.

Frame Relay

Frame Relay was introduced in 1992, and has since grown as a popular alternative to PPP, ISDN, X.25, and other WAN technologies. Frame Relay assumes a high-speed link and is able to transmit faster than its predecessor X.25 because it does not have error correction.

WANs and Internet Access

The advantages of Frame Relay are:

- It's widely adopted in the U.S. and internationally.
- Its high-capacity switched core offers scalability.
- It's a flexible technology that allows for expansion and growth.

The disadvantages of Frame Relay are:

- It provides minimal error correction.
- Connection-oriented service can cause latency on slower links.
- It's expensive ($350–$1500 per month, depending on service).

Integrated Service Digital Network (ISDN)

In the 1980s, ISDN was believed to be the solution for supplying residential sites and small businesses with an integrated data and voice solution. Unfortunately, many factors have prevented the wide adoption of ISDN that telecommunications companies once thought would be possible. In particular, ISDN has been overrun by more affordable solutions that offer even greater bandwidth potential at lower prices.

As you learned in Chapter 10, ISDN comes in two types: Basic Rate Interface (BRI) and Primary Rate Interface (PRI). BRI supports up to 128Kbps on two B channels of 64Kbps each and a 16Kbps D channel. Either B channel can be used for voice, data, or a combination of the two. PRI is found in much larger installations. PRI has 23 B channels and one 64Kbps D channel.

It should be noted that although ISDN did not make telephone companies much money in the past, they have found a use for ISDN. Many companies have repackaged ISDN as IDSL. It's not DSL technology, but it does allow DSL providers the ability to offer access to customers that are more than 1500 feet from the CO.

The advantages of ISDN are:

- It runs on existing telephone lines.
- It offers support for data, voice, and fax on one line.
- It's excellent for video conferencing.

The disadvantages of ISDN are:

- It's expensive ($75–$150 and up per month for BRI, depending on usage time) unless it is packaged as IDSL, in which case it is a flat rate.
- It doesn't provide much bandwidth (if no voice channels are in use, the bandwidth is 128Kbps).

- It provides less flexibility for growth, and costs more than other WAN technologies.
- ISDN service is not available everywhere.
- It can be difficult to implement.

Avoiding the Potholes

Before you make your final decision about your WAN connection, take the time to do some research and planning. The first item you should consider is what the total cost for your WAN connection will be. When you contact an ISP for a quote on a T1 connection, the first price you are quoted might sound like a good deal. (Clear-channel T1s start under $500 per month in San Francisco!) But this price may only be the Internet provider's monthly charge. There are several other charges you should be aware of:

- The ISP setup fee
- Hardware costs for routers, CSU/DSU, DSL modems, and the like
- The telephone company setup fee
- The telephone company monthly line charge

Outside of these costs, many ISPs offer additional services as part of the package, including:

- Additional IP addresses
- Domain name services for hosting websites and e-mail servers
- Space on an ISP web server for your website
- Free e-mail accounts, with the ability to add more later

Take advantage of these services whenever possible. Not only are you paying for them, but the ISP equipment is usually much faster than what a small- or medium-sized business could afford. Also, the ISP usually has a highly trained staff available most hours to make sure the network is up and running. Be sure to inquire about support hours when selecting your ISP.

The Rollout

After you have found the ISP you are going to use and selected the service you want, you will need to decide on a date to "go live" with your new connection. If this is your company's first Internet connection, the rollout should go smoothly because end users who don't already have access won't have many expectations.

WANs and Internet Access

When making the transition from an existing Internet connection to a new Internet provider, however, problems are more likely to occur. The following list of goals for the transition should steer you in the right direction:

- There should be 99 percent uptime in Internet access for users.
- There should be no downtime apparent to external users, especially clients.
- All services should be moved to the new ISP.
- Domain names resolve successfully for all e-mail, web, and other servers.

Actually meeting your goals for the transition isn't always easy. This type of transition is a major task that only becomes more complicated with larger companies. To be successful, you will need to have a solid game plan.

Here are some suggestions for your plan:

- Identify the main technical contacts for both the old and the new ISPs, including phone numbers and e-mail addresses. Make sure to include the contact information of the DNS administrators at both ISPs.
- Have conversations with the technical contact at the new ISP, and coordinate the transition with that person.
- Determine how long your **domain name registry** company, also known as the **registrar**, is taking to fulfill domain name registrations and modifications.
- Make sure all equipment is ordered, and check for delivery dates.
- Confirm the date the telephone company will install the line.
- After the line is installed, confirm that the ISP's line is up and that they are online.
- Request the IP addresses and plan the deployment of the new addresses.

Domain Name Registry
Any authorized organization that is responsible for the management of domain names.

TIP

During a transition from one ISP to another, the IP addresses will have to change. The IP addresses belong to the ISP. To avoid the difficulties of IP address planning, consider implementing DHCP (discussed in Chapter 11).

- Install new equipment at your site, and confirm that there is a connection with the ISP.
- Submit domain name registration or modification to your registrar, and copy the technical contact and DNS administrator at the new ISP. There is no telling how long it will take to update the root (main) DNS servers.

The process usually takes less than 24 hours, but it is not unheard of for the process to take as long as 3 or 4 days. It is a good idea to have your old ISP add the new IP address of your e-mail or web server to its DNS servers. In case someone does request the IP address of your server from your old ISP, they will get the new IP address.

- Make the transition in the middle of the night on the same day that the registration or modification is submitted to InterNIC.

After the WAN connection is up and people are using the Internet, the fun really begins! Good luck!

Review Questions

1. What steps should be taken before researching Internet access solutions?

   _____

   _____

2. What telecommunications equipment is located between a home DSL Internet connection and the Internet service provider?

   _____

   _____

3. What is multiplexing?

   _____

   _____

4. What is a multichannel trunk?

   _____

   _____

5. What is the total capacity of a T1 line? How much is available for data?

   _____

   _____

6. What are two advantages of using cable modems for Internet access?

   _____

   _____

Terms to Know
- ❏ cell switching
- ❏ demultiplexed
- ❏ dial-up
- ❏ DS0
- ❏ e-commerce
- ❏ Domain Name Registry (registrar)
- ❏ leased line
- ❏ local exchange carrier (LEC)

Review Questions

Terms to Know
- local loop
- metropolitan exchange
- multichannel trunk
- multiplexed
- premise
- subscriber
- switched circuits
- T-carrier

7. What additional costs are involved with Internet access?

8. Why is DSL popular?

9. How much downtime is acceptable when changing Internet service providers?
   A. No more than half a day
   B. One full day
   C. None
   D. One percent or less

10. When changing ISPs, why do you need to contact your domain registry company (registrar)?

11. Why do you have to reconfigure the IP addresses for your company when you transfer to a new ISP?

Appendix A

Answers to Review Questions

Chapter 1

1. Based on AT&T's proposal, the government believed that giving AT&T the monopoly over the local phone service was the simplest way for telephone access to be provided to all homes in the United States.

2. The 1996 ruling in favor of the Carterphone allowed anyone the right to attach a device to the telephone network, provided the device would not damage the network. The case made it possible for other telephone manufacturers and later modem manufactures to attach their equipment to the telephone network.

3. The Department of Defense developed ARPAnet.

4. Robert Metcalf created the first Ethernet network.

5. The four main characteristics of LANs are as follows:

    They're installed in a small area, such as an office building or floor.

    They're high-speed networks.

    Devices have equal access to the network.

    They use LAN-specific equipment such as repeaters, hubs, and network interface cards.

6. MANs are used in situations in which multiple office locations in the same metropolitan area need to be connected.

7. WAN connections link LANs and MANs together over large geographical areas at much slower speeds than LANs or MANs provide.

8. Two of the reasons ARPAnet was successful are as follows:

    The original design of ARPAnet called for a system that could still maintain communication in the event of a partial failure.

    The distribution of BSD UNIX gave universities an operating system to connect to ARPAnet.

9. The National Science Foundation's NSFNET provided a fast network backbone for the Internet, increasing the number of networks connected to the Internet and its overall capacity.

10. The major accomplishments of Internet 2 are as follows:

    Internet 2 provides institutions and researchers exclusive access to one of the fastest networks connecting multiple educational, governmental, and research institutions.

    Internet 2 runs at gigabits per second.

Answers to Review Questions

Chapter 2

1. The purpose of the OSI model is to provide a theoretical model for intercommunication between networking devices.

2. The seven layers of the OSI model are as follows:
    Layer 1: Physical layer
    Layer 2: Data Link layer
    Layer 3: Network layer
    Layer 4: Transport layer
    Layer 5: Session layer
    Layer 6: Presentation layer
    Layer 7: Application layer

3. Each layer provides services to the layers above and below it while maintaining the same functions as the corresponding layer of other network devices.

4. The functions provided at each layer include:
    Layer 7: To provide file, mail, or printing services to the end user
    Layer 6: Data presentation; data compression; data encryption
    Layer 5: To establish, maintain, or end a communication; dialog control; dialog separation
    Layer 4: Connection-oriented and connectionless transmissions; flow control
    Layer 3: Select the correct path to reach the destination; encapsulation of the network address
    Layer 2: Determine whether connection-oriented or connectionless; encapsulate the packet with the MAC address
    Layer 1: Convert data to binary form; transmit the signal over the medium

5. The Session layer is responsible for the setup and breakdown of a communication.

6. The Network layer is responsible for end-to-end routing of the data.

7. The physical address of the device is added to the data packet at the Data Link layer.

8. The physical address of a device is called the he MAC address. The manufacturer hard codes the address into the NIC.

9. The network address is added at the Network layer.
10. Encapsulation is the process of adding an address.
11. The two sublayers of the Data Link layer are the Logical Link Control (LLC) sublayer and the Media Access Control (MAC) sublayer.
12. The two types of communications used on a network are connectionless and connection-oriented.
13. Twisted-pair cable, fiber-optic cable, and coaxial cable are all types of physical connections that can be used in networks.
14. Some devices that function at the Physical layer include:

    Repeater

    Hub

    NIC

    Concentrator

Chapter 3

1. Peer-to-peer, client-server, and hybrid are examples of network architectures.
2. Some characteristics of peer-to-peer networks include:

    No server hardware or software needed

    Used in small networks of 10 users or fewer

    Management of resources is maintained by each user

    No centralization of resources or management

    Poor security

3. There is minimal security on a peer-to-peer network. Security is maintained by each individual user, which allows for inconsistencies in security policies. Typically, only one password can be assigned per resource.
4. A server is a computer used to centralize resources, administration, and security using specialized hardware and software.
5. The three hardware components needed to connect computers in a peer-to-peer network are hub, cable, and network interface card (NIC).

Answers to Review Questions

6. Network configuration is performed through the Network Control Panel. You can get to the same configuration window by right mouse clicking on My Network Places on the Desktop.

7. Folders and printers have sharing options that can be accessed by viewing the Properties window for that item. Before the folder or printer can be shared, though, a user account needs to be created. The user account is created using the Users and Passwords control panel.

8. Some disadvantages of peer-to-peer networks include:

    Poor security

    No central location for file saving

    Limited to 10 users or fewer

    Limited number of connections to share resources

    Poor performance under heavy load

    No centralized management

9. Some advantages of client-server networks include:

    High level of security

    Data stored centrally and easily backed up

    Centralized management

    Improved performance

    Users relieved of network management responsibilities

    Most powerful network resources available to all users

10. The distinction is as follows:

    A file server centrally stores user data. The data is secured for each user using access rights. The file server can make files and folders available to hundreds of users while maintaining security.

    The messaging server centralizes the exchange of electronic mail. The client requests to receive or send mail information that is stored on the messaging server.

11. Some factors in deciding whether a single-server, multiserver, or enterprise network is needed include:

    Number of users

    Application needs

    Geographical locations of users

    Amount of tech support available

335

Chapter 4

1. Multitasking (and pre-emptive multitasking) make it possible for the CPU to perform many tasks simultaneously.

2. Multiprocessing is a function of the network operating system to manage two or more CPUs to complete a task or several tasks.

3. You need UPS, RAID, and a tape backup drive to create a fault-tolerant server.

4. The Domain controller is a Windows 2000 server that maintains a complete copy of the Active Directory database. The Active Directory database stores all information, security and otherwise, about users, groups, and resources available on the network.

5. NDS is a software management system that allows a network administrator to organize users and devices—including NetWare, Windows 2000, Macintosh OS X and UNIX servers—on the network.

6. Some advantages of the UNIX operating system include:

    Stable operating system built on more than 30 years of use

    Quick processing of requests

    Functions as both server and workstation

    Supports multiple users

    Many tools available for supporting, using, and administering UNIX

7. There's a lot of interest in Linux because it runs on PCs using Intel 386 or faster processors, it's fast like UNIX, and the operating system and many applications are free.

8. Some of the reasons you might want to consider using a MacOS X server include:

    MacOS X is built on FreeBSD, so it has all the flexibility of a Linux system.

    Unlike some Linux software, Apple supports MacOS X as an integrated solution on Apple hardware.

    MacOS X runs the Apache web server, which is the most widely used web server on the Internet.

9. The ipconfig command displays the IP address, the adapter address, the subnet mask, and the gateway. By using the command ipconfig /all the DNS, DHCP and additional server information is displayed.

Answers to Review Questions

10. The distinction is as follows:

    Ping is an application that is used to test connectivity between two computers configured with IP addresses.

    Traceroute (the Tracert command in Windows XP) gives details about the path a packet takes trying to reach the destination address.

Chapter 5

1. The physical topology is the physical layout of the network media.
2. The logical topology is the way in which data is transmitted throughout the network.
3. The most common physical topologies are bus, ring, star, mesh, and hybrid.
4. The physical star topology is the most common physical topology today.
5. Thin coaxial cable, or Thinnet, is typically used with a physical bus topology.
6. Ethernet is usually used to implement a physical bus topology.
7. The advantages of installing a physical bus topology are as follows:

    It's inexpensive to install.
    More workstations are easily added.
    Uses less cable than other physical topologies.
    Works well for small networks (2–10 devices).

    The disadvantages of installing a physical bus topology are as follows:

    Not recommended for new installations.
    If the backbone breaks, the network is down.
    Only a limited number of devices can be included.
    Problems are difficult to isolate.
    Slower access time.

8. A physical ring topology connects all devices, one to the next, in the form of a closed ring; all devices are connected to two others to ensure the completion of the ring.
9. Token Ring and FDDI both use a physical ring topology.
10. The advantages of installing a physical ring topology are as follows:

    Can pass data packets at greater speeds.
    No collisions.

Easier to locate problems with devices and cable.

No terminators needed.

The disadvantages of installing a physical ring topology are as follows:

Requires more cable than a bus network.

A break in the cable will bring down many ring networks.

Adding devices to the ring suspends all devices from using the network.

Not as much equipment available.

11. In a physical star topology, each device is connected to a central device, known as a hub, creating a star.

12. Ethernet and Token Ring can be implemented using a physical star topology.

13. In a mesh topology, all devices are physically connected to each other.

14. The advantage of installing a physical mesh topology is:

The network has fault tolerance in case of device or cable failure.

The disadvantages of installing a physical mesh topology include:

Expensive and difficult to install

Difficult to manage

Difficult to troubleshoot

15. The two logical topologies used in networks are bus and ring.

16. The logical topologies that can be implemented with a physical star are bus and ring.

17. Physical bus, physical star, and logical bus can be implemented with Ethernet.

18. Physical ring, physical star, and logical ring can be implemented with Token Ring.

19. There are no physical limitations of wireless LANs, other than interference caused by walls or other structures. Wireless LAN transmissions have different distance limitations, depending on the wireless standard they follow.

20. Ad-hoc, or peer-to-peer mode is manner of transmitting wireless signals between devices or stations *without* the use of an access point.

21. Infrastructure mode is a manner of transmitting wireless signals that utilizes an Access Point as an extension of a wired network. This allows communication between wireless nodes or the wired network.

Answers to Review Questions

Chapter 6

1. The two types of electrical currents we use to provide power to our devices are alternating current (AC) and direct current (DC).

2. Alternating current is current whose voltage rises and falls due to the flow of electrons from positive to negative.

3. Direct current is current whose voltage remains constant.

4. Static electricity is created by the friction caused by our movements.

5. If an electrostatic charge that builds up in a person's body is displaced into electronic equipment, it can damage the equipment, especially the microchips inside a computer.

6. To prevent an electrostatic discharge, wear a properly grounded wrist strap when working on devices, and displace static build-up in your body by touching a grounded metal device such as a computer cabinet.

7. Surges and spikes can result from excess electrical current in the line.

8. A surge is an increase in the electrical signal to 110 percent or more of the normal voltage for a short amount of time.

9. A spike is an increase in the electrical signal of at least twice the normal voltage for 10 to 100 microseconds.

10. To prevent excess electrical current from damaging your devices, install surge suppressors for all devices.

11. Sags, brownouts, and blackouts can result from a reduction in electrical signal.

12. A sag or brownout is a reduction of electrical current to 80 percent of the normal voltage or less.

13. A blackout is a complete loss of electrical signal.

14. Use uninterruptible power supply (UPS) devices to prevent a loss of data or damage to devices as a result of sags or blackouts.

15. A power conditioner, or line conditioner, is a device that is installed between a building's power source and the devices being protected. It increases or decreases voltage to regulate the power that is being delivered.

16. Some of the special considerations that should be addressed when planning a data center include power requirements, adequate cooling, sufficient space, and raised flooring.

Chapter 7

1. A digital signal makes a square wave.
2. The number of cycles per second is measured in hertz.
3. The speeds match up as follows:

    Mbps = 1,000,000 bits per second
    Kbps = 1,000 bits per second
    Gbps = 1,000,000,000 bits per second
    bps = 1 bit per second

4. When a series of bits is transmitted as a digital signal, it is called a bit stream.
5. It's important to have a clear digital signal because transmissions that are not precisely the right value (e.g. a "1") may be misinterpreted as the other value (e.g. a "0") and can cause errors in transmission.
6. Modems convert digital signals into analog signals that can be sent over telephone lines. Modems allow people to access the Internet and other computers using the telephone system, which is widely available throughout the world.
7. Answer:

Symmetrical	Data is transmitted in blocks of bits.
Symmetrical	Performance is not as good on poor-quality lines.
Asymmetrical	It's less efficient due to overhead of extra bits.
Symmetrical	It transmits faster because of no excess overhead.
Asymmetrical	It's best suited to traffic that occurs in short bursts.

8. Half duplex allows data to be transmitted in both directions, but only one way at a time.

Chapter 8

1. The three types of transmission media for networks are copper, glass, and air.
2. Coaxial cable, shielded twisted pair, and unshielded twisted pair are all copper media.
3. Thicknet is difficult to work with and expensive.

Answers to Review Questions

4. Some advantages to using Thinnet are as follows:

    Easy to implement
    Cheaper than Thicknet
    Reduced interference
    Small in diameter

5. The distance limitation for a Thinnet segment is 185 meters.

6. BNC connectors are used with Thinnet.

7. STP is typically found in IBM Token Ring environments.

8. Some advantages to STP are as follows:

    Reduced cross talk
    Less susceptible to EMI and RFI due to the use of foil or braided shielding
    Can be used with RJ connectors

9. The most common unshielded twisted-pair cable used today is Category 5.

10. Some disadvantages of Cat 5 UTP are as follows:

    Susceptible to interference
    Can be easily damaged during installation
    Can cover only a limited distance

11. Four pairs of twisted color-coded wires are found in Cat 5e UTP.

12. Cat 6 was designed to support applications running at one gigabit per second or better.

13. The connectors used for UTP are called RJ connectors.

14. Single-mode and multimode are two types of fiber-optic cable.

15. The distance limitation for multimode fiber is two kilometers per segment.

16. The three types of wireless transmission are radio, microwave, and infrared.

17. Some advantages of wireless transmission include:

    You don't have to install physical media.
    They can cover greater distances than physical media can.

18. The distinction is as follows:

    Satellite microwave transmissions use satellites that orbit the earth, and can be accessed from anywhere in the world.
    Terrestrial microwave transmissions use line-of-sight dishes to exchange data communications, and have a limit of 50 miles.

Chapter 9

1. The network interface card, or NIC, is responsible for the connection between a device and the network media.
2. The MAC (Media Access Control) address is a hexadecimal number that identifies the physical address of the NIC card; it is hard-coded into the NIC by the manufacturer.
3. The NIC functions at the Data Link layer.
4. Repeaters and hubs function at the Physical layer.
5. Hubs take in signals and pass them to all other connected devices. They do not filter any traffic, and pass on all collisions and broadcasts.
6. A collision domain is an area of the network, bound by bridges, switches, or routers, in which collisions are propagated.
7. A network segment is bound by bridges and switches that prevent the propagation of collisions and by routers that prevent the propagation of broadcasts.
8. An access point is a device that connects wireless users to a network.
9. A bridge makes a decision to forward data or drop it, based on the destination MAC address.
10. A wireless bridge is a device that connects wired LAN segments using wireless transmissions.
11. A switch is a multiport bridge. It makes the decision to forward data or drop it, based on the destination MAC address.
12. Virtual LANs allow network segments to be created using software. Each VLAN that is created is its own broadcast domain.
13. Bridges and switches function at the Data Link layer.
14. A router routes data packets to distant networks via the best route, based on the network address.
15. Routing tables enable routers to make decisions about packet forwarding.
16. Two types of routing are static and dynamic.
17. Static routes are determined by the network administrator.
18. Dynamic routes are determined by the router, based on information from neighboring routers.

Answers to Review Questions

19. A gateway functions as a translation device to allow dissimilar networks to communicate.

20. Gateways can function at all layers of the OSI model.

Chapter 10

1. Standards define the rules that allow interoperation between devices, equipment, and media.

2. The six primary organizations that adopt networking standards are as follows:

    International Organization for Standardization (ISO)

    Institute of Electrical and Electronics Engineers (IEEE)

    Electronics Industries Alliance (EIA)

    Telecommunications Industry Association (TIA)

    American National Standards Institute (ANSI)

    International Telecommunications Union (ITU)

3. The ISO is the International Organization for Standardization. It developed and adopted the OSI model.

4. 802.3 is the Carrier Sense Multiple Access/Collision Detection (CSMA/CD) access method, better known as Ethernet. It requires that all devices listen before transmitting data across a network and that a backoff be issued in the event of a collision.

5. A backoff is the forced termination of data transmissions on an Ethernet segment after a collision.

6. 802.3u is Fast Ethernet—a high-speed, data-transmission standard that uses CSMA/CD to transmit at 100Mbps.

7. 802.3z is Gigabit Ethernet; it transmits at a rate of one billion bits per second.

8. Gigbabet Ethernet uses CSMA/CD, Carrier Sense Multiple Access/Collision Detection.

9. The 802.11 standards support wireless LAN technologies.

10. 802.11b is the most common wireless LAN standard today.

11. Although currently still in development, 802.15, the Wireless Personal Area Network standard, will support connectivity between most mobile devices when it is adopted.

Appendix A

12. 802.5 is the Token Ring access method. It uses a token-passing for devices to transmit data.
13. The EIA/TIA is the combination of two standards organizations—the Electronics Industry Alliance and the Telecommunications Industry Association.
14. 569 defines the size, location, and number of wiring closets that support the telecommunications infrastructure of a commercial building.
15. The 602 standard is important because without proper grounding, the data and devices can be damaged or destroyed by electrical surges.
16. PSTN stands for the public switched telephone network.
17. The only internationally adopted WAN protocol is the X.25 packet-switching protocol.
18. ATM stands for Asynchronous Transfer Mode.
19. FDDI uses fiber-optic cables that incorporate dual rotating rings to pass tokens at high speed over a logical ring network.

Chapter 11

1. A protocol is a set of formal rules that defines how devices on a network exchange information.
2. A protocol suite is a set of protocols that work together at different layers to provide complete end-to-end communication.
3. Four layers are represented in the TCP/IP protocol suite: Application layer, Transport layer, Internet layer, and Data Link layer.
4. IP is responsible for logical network addressing to route information between networks.
5. A subnet mask allows a network administrator to segment a large network into smaller subnetworks.
6. ARP sends out a MAC broadcast packet, asking for the MAC address of the device with the known IP address.
7. RARP is used when diskless workstations need to be able to request IP addresses to communicate on the network.
8. TCP provides connection-oriented services to guarantee delivery.
9. UDP uses connectionless delivery, and assumes that other layers will provide the error checking.

Answers to Review Questions

10. FTP allows files to be transferred and copied between devices.
11. SMTP allows e-mail to be transferred between computers using connection-oriented services.
12. DNS translates the domain name, for example a website, such as www.sybex.com, into an IP address.
13. DHCP gives IP addresses to devices that want to communicate on the network for a specified amount of time.
14. SNMP is an Application layer protocol that can be used to configure, monitor, and manage devices on the networks that understand the SNMP protocol.
15. AppleTalk is considered the "chattiest" protocol. AppleTalk broadcasts occur every 10 seconds, consuming a large portion of available bandwidth.
16. The first four bytes of an IPX address are the network address, which is assigned by the network administrator. The MAC address of the device is used for the next six bytes of the IPX address. Finally, two bytes are used for the IPX socket number.
17. ZIP is responsible for keeping track of network numbers and zones. It matches the network number to the appropriate zone within an AppleTalk network.

Chapter 12

1. The phases of a successful LAN design are as follows: needs assessment, design; development, implementation, and evaluation.
2. During a needs assessment, you should document user needs, the building layout, and office locations; and you should take an inventory of all equipment.
3. The equipment inventory provides a picture of potential performance problems related to the hardware capabilities of devices.
4. A facility assessment should include the size of the building offices; the position of furniture; and the location of electrical outlets, phone jacks, the POE for telephone lines, ventilation shafts, lighting, and plumbing.
5. If the company has a dedicated network staff and wants to maintain system-wide security on all computers and minimize support calls, you should choose client-server architecture.
6. IDFs are needed in multistory buildings and large office buildings (where the distance from the MDF to any office is beyond a 65-meter radius).

7. Before you utilize wireless networks, it is important to determine which structures in the building will impede wireless performance. The obstacles can include concrete walls, aluminum studs, and steel beams. In addition, electromagnetic fields from other wireless devices or large electrical systems can cause significant interference.

8. A distribution-layer device provides services to access-layer devices. An access-layer device provides connections directly to the computers.

9. A TDR cable tester provides details about the length of the cable, and pinpoints the location of breaks in the cable; some will test for near-end cross talk and the level of attenuation on the line.

10. You would use a router on your LAN to break up the LAN into smaller network segments to improve performance or to connect the LAN to a WAN.

11. When coordinating Internet access, the network administrator needs to contact an ISP and the telephone company.

Chapter 13

1. Fault tolerance can be supplied to a network using redundancy, tape backup, and having a universal power supply (UPS).

2. In most organizations, Level 1 support personnel have been trained to solve the most common types of issues that users experience with their computers. The issues usually involve problems using desktop applications and peripherals.

3. Remote desktop control applications make it possible for support staff to troubleshoot problems remotely without the need to go directly to the user's computer. This can save precious time, especially in larger companies that have multiple buildings or locations.

4. Travan drives are a more cost effective solution for small businesses. Although Travan drives are slower and have a lower storage capacity than DAT drives, they are a proven and reliable low cost alternative..

5. A differential backup copies files that have changed since the last normal backup. Differential backups take longer, but only one tape out of the set is needed to restore data. An incremental backup copies files that have changed since the last incremental backup. Incremental backups are faster, but take longer to restore because several incremental tapes need to be used to recover all files.

6. Using a performance-monitoring application; using a network management console (if the server supports SNMP).

Answers to Review Questions

7. The type of traffic; the protocols most frequently used; the number of collisions; the devices transmitting the most packets; the frequency of packet errors.

8. The MIB file contains the definitions and values for the objects that are specific to the managed device. The MIB file can provide this information to the agent to be used in response to SNMP commands from the management console.

9. Journal; problem log; books; vendor documentation; software update information.

Chapter 14

1. The steps to be taken before researching Internet access solutions are as follows: identify user needs; recognize possible future applications, and determine current WAN infrastructure.

2. The telecommunications equipment between the user and the ISP are the following: DSL router or modem, local loop (telephone wire), CO exchange (switch), trunk line to the ISP, and the ISP's router.

3. Multiplexing is sending multiple voice or data signals over a single wire.

4. A multichannel trunk is a dedicated leased line of four wires that is capable of supporting 24 DS0 or voice channels. Each channel can transmit at up to 64Kbps.

5. The total capacity of a T1 line is 1.544Mbps, with 1.536Mbps (8Kbps used for framing) available for data.

6. Some advantages of using cable modems include the following:

    Data can be transmitted at speeds up to 10Mbps.

    Existing television cable is used.

    Television transmission and data transmission are available on the same line simultaneously.

    There is future potential for telephone service over cable.

    Costs less than $50 per month for most services.

7. The additional costs involved with Internet access include the following: ISP setup fee, equipment costs, installation charge for telephone line or cable, and monthly telephone line charge.

Appendix A

8. DSL provides high-speed Internet access from 128Kbps up to 6Mbps. DSL is inexpensive, runs on normal phone lines, and can transmit voice and data on the same physical line. DSL is widely available across the United States.

9. D. 1 percent or less should be the acceptable downtime for changing ISPs.

10. You need to notify them that your domain name will be hosted by a different ISP.

11. IP addresses belong to the ISP. When switching to a new ISP, it is necessary to relinquish old IP addresses and use the ones provided by the new ISP.

Appendix B

Acronyms and Abbreviations

Appendix B

Acronym or Abbreviation	What It Stands For
AARP	AppleTalk Address Resolution Protocol
AC	alternating current
ack	acknowledgment
ACL	access control list
ACP	access control point
ADB	Apple Desktop Bus
ADC	analog-to-digital converter
ADSI	Analog Display Services Interface
ADSL	Asymmetrical Digital Subscriber Line
ADSP	AppleTalk Data Stream Protocol
AEP	AppleTalk Echo Protocol
AF	address field
AFI	AppleTalk Filing Interface
AFP	AppleTalk Filing Protocol
AFT	Application File Transfer
AIO	asynchronous input/output
AL	Application layer
AM	amplitude modulation
AMH	Application Message Handling
ANS	American National Standard
ANSA	Advanced Network Systems Architecture
ANSC	American National Standards Committee
ANSI	American National Standards Institute
AOL	America Online
AP	access point
APCI	Application Layer Protocol Control Information
APDU	Application Protocol Data Unit
API	Application Program Interface
ARAP	AppleTalk Remote Access Protocol
ARM	Asynchronous Response Mode
ARP	Address Resolution Protocol

Acronyms and Abbreviations

Acronym or Abbreviation	What It Stands For
ARPA	Advanced Research Projects Agency
ARPAnet	Advanced Research Projects Agency network
AS	application system
ASB	Asynchronous Balanced Mode
ASCII	American Standard Code for Information Interchange
ASDU	Application layer Service Data Unit
ASE	Application Service Element
ASO	Application Service Object
ASP	AppleTalk Session Protocol
async	asynchronous
AT	advanced technology
AT&T	American Telephone and Telegraph
ATM	Asynchronous Transfer Mode
ATM-SDU	Asynchronous Transfer Mode Service Data Unit
ATP	AppleTalk Transfer Protocol
ATPS	AppleTalk Print Services
ATQ	AppleTalk Transition Queue
BCP	Byte Control Protocol
BCVT	Basic Class Virtual Terminal
BDT	Bureau of Telecommunications Development
BER	bit error rate
BGP	Border Gateway Protocol
BIOS	Basic Input/Output System
BISDN	Broadband ISDN
bit	binary digit
BLNT	Broadband Local Network Technology
BNC	British Naval Connector *or* Bayonet Neill-Concelman
BNT	broadband network termination
BOOTP	Bootstrap Protocol
bps	bits per second

Appendix B

Acronym or Abbreviation	What It Stands For
Bps	bytes per second
BRI	Basic Rate Interface
BTE	broadband terminal equipment
BTR	bit transfer rate
CAI	computer-aided instruction
CAN	campus area network
CAU	Controlled Access Unit
CBR	constant bit rate
CCDA	Cisco Certified Design Associate
CCDP	Cisco Certified Design Professional
CCIA	Computer and Communications Industry Association
CCIE	Cisco Certified Internetworking Expert
CCITT	Consultative Committee for International Telephony and Telegraphy
CCNA	Cisco Certified Network Associate
CCNP	Cisco Certified Network Professional
CCR	current cell rate
CDDT	Copper Distributed Data Interface
CET	Computer-Enhanced Telephony
CHAP	Challenge Handshake Authentication Protocol
CIDR	Classless Interdomain Routing
CIPX	Compressed Internet Packet eXchange
CLTP	Connectionless Transport Protocol
CMIP	Common Management Information Protocol
CMOT	Common Management information protocol Over TCP/IP
CNE	Certified NetWare Engineer
CNOS	computer network operating system
CO	central office
coax	coaxial cable
COS	Class of Service

Acronyms and Abbreviations

Acronym or Abbreviation	What It Stands For
CPU	central processing unit
CRC	cyclic redundancy check
CRS	Cell Relay Service
CS	circuit switching
CSDN	Circuit Switched Data Network
CSMA/CA	Carrier Sense Multiple Access/Collision Avoidance
CSMA/CD	Carrier Sense Multiple Access/Collision Detection
CSMA/CP	Carrier Sense Multiple Access/Collision Prevention
CSO	central switching office
CSU	channel service unit
DAC	digital-to-analog converter
DAF	destination address field
DAK	data acknowledge
DAN	departmental area network
DAP	Directory Access Protocol
DARPA	Defense Advanced Research Projects Agency
DAS	Disk Array Subsystem
DAT	dynamic address translation *or* digital audio tape
DC	direct current
DCA	Defense Communications Agency
DCE	data circuit-terminating equipment
DCP	Digital Communications Protocol
DCS	data circuit switches
DDP	Datagram Delivery Protocol
DDR	Dial on Demand Routing
DDS	digital data service
DEC	Digital Equipment Corporation
DECnet	Digital Equipment Corporation network
DEMUX	demultiplexer
DFC	data flow control
DFS	Distributed File System

Appendix B

Acronym or Abbreviation	What It Stands For
DHA	destination hardware address
DHCP	Dynamic Host Resolution Protocol
DIT	directory information tree
DL	distribution list
DLC	Data Link Control
DLL	Data Link layer
DLPDU	Data Link Protocol Data Unit
DMSP	Distributed Mail System Protocol
DMUX	double multiplexer
DNS	Domain Name System
DoD	Department of Defense
DOV	data over voice
DP	demarcation point
DSAP	destination service access point
DSC	direct satellite communications
DSE	data-switching equipment
DSL	Digital Subscriber Line
DOCSIS	Data Over Cable System Interface Specification
DSU	data service unit
DSU/CSU	data service unit/channel service unit
DTE	data terminal equipment
DU	data unit
EBCDIC	Extended Binary Coded Decimal Interchange Code
EDAC	error detection and correction
EDDP	Extended Datagram Delivery Protocol
EFS	external file system
EIA	Electronics Industries Alliance
EIGRP	Enhanced Internet Gateway Routing Protocol
ELAP	EtherTalk Link Access Protocol
EMI	electromagnetic interference
EP	Echo Protocol

Acronyms and Abbreviations

Acronym or Abbreviation	What It Stands For
ESD	electrostatic discharge
ES-ES	end system to end system
ES-IS	end system to intermediate system
ESMTP	Extended Simple Mail Transfer Protocol
FAQ	frequently asked questions
FAT	File Allocation Table
fax	facsimile
FCC	Federal Communications Commission
FDDI	Fiber Distributed Data Interface
FDM	frequency-division multiplexing
FDMA	Frequency Division Multiple Access
FDX	full duplex
FLAP	FDDI Link Access Protocol
FLIP	Fast Local Internet Protocol
FO	fiber optics
FOC	fiber-optic communication
FOIRL	Fiber Optic Inter-Repeater Link
fps	frames per second
FPSNW	File and Print Services for NetWare
FQDN	fully qualified domain name
FR	Frame Relay
FS	file system
FT	fault tolerant
FTP	File Transfer Protocol
FTS	File Transfer Service
FTSC	Federal Telecommunications Standards Committee
GAN	global area network
GAP	Gateway Access Protocol
Gb	gigabit
GB	gigabyte
Gbps	gigabits per second

Appendix B

Acronym or Abbreviation	What It Stands For
GBps	gigabytes per second
GFC	Generic Flow Control
GGP	Gateway-to-Gateway Protocol
GHz	gigahertz
GMHS	Global Message Handling Service
GSN	Government Satellite Network
GTA	Government Telecommunications Agency
GUI	graphical user interface
HAND	have a nice day
HCC	horizontal cross-connect
HCL	hop count limit *or* hardware compatibility list
HDLC	High-Level Data Link Control
HDSL	High-Bit-Rate Digital Subscriber Line
HDTV	high-definition television
HDX	half duplex
HLPI	Higher-Level Protocol Identifier
HPAD	Host Packet Assembler/Disassembler
HPFS	High-Performance File System
HRC	Hybrid Ring Control
HSLAN	high-speed local area network
HTML	HyperText Markup Language
HTTP	HyperText Transfer Protocol
Hz	hertz
I/O	input/output
IAN	integrated analog network
IBM	International Business Machines
IC	integrated circuit
ICMP	Internet Control Message Protocol
ICP	Internet Control Protocol
IDE	Integrated Drive Electronics
IDF	intermediate distribution facility

Acronyms and Abbreviations

Acronym or Abbreviation	What It Stands For
IDN	integrated digital network
IDP	Internet Datagram Protocol
IDPR	Interdomain Routing Protocol
IEEE	Institute of Electrical and Electronics Engineers
IGP	Internet Gateway Protocol
IGRP	Internet Gateway Routing Protocol
IMAP	Internet Message Access Protocol
IMP	information management plan
InARP	Inverse Address Resolution Protocol
IOS	Internetwork Operating System *or* Intermediate Open System
IP	Internet Protocol
IPng	Internet Protocol, next generation
IPX	Internetwork Packet eXchange
IPX/SPX	Internetwork Packet eXchange/Sequence Packet eXchange
IR	infrared
IRQ	interrupt request
ISDN	Integrated Services Digital Network
ISO	International Organization for Standardization
ISP	Internet service provider
ISR	Intermediate Session Routing
ITU	International Telecommunications Union
ITU-T	International Telecommunications Union-Telecommunications
IVD	integrated voice and data
IVOD	integrated voice on demand
Kb	kilobit
KB	kilobyte
Kbps	kilobits per second
KBps	kilobytes per second
KHz	kilohertz

Appendix B

Acronym or Abbreviation	What It Stands For
KMP	Key Management Protocol
LAN	local area network
LAPB	Link Access Procedure, Balanced
LAPD	Link Access Procedure, D Channel
LAT	Local Area Transport
LAWN	local area wireless network
LCP	Link Control Protocol
LCR	Least Cost Routing
LDAP	Lightweight Directory Access Protocol
LED	light-emitting diode
LLC	Logical Link Control
LLC1	Logical Link Control type 1
LLC2	Logical Link Control type 2
LLP	lower-layer protocol
LLPDU	lower-layer protocol data unit
LOS	line of site
MAC	Media Access Control
MAN	metropolitan area network
MAPI	Mail Application Program Interface
MAU	Multistation Access Unit
Mb	megabit
MB	megabyte
Mbps	megabits per second
MBps	megabytes per second
MCP	MAC Convergence Protocol *or* Microsoft Certified Professional
MCSE	Microsoft Certified Systems Engineer
MDF	main distribution facility
MDS	Mail Delivery System
MFS	Macintosh File System
MHP	Message Handling Protocol

Acronyms and Abbreviations

Acronym or Abbreviation	What It Stands For
MHz	megahertz
MIB	management information base
MIT	Massachusetts Institute of Technology
MMF	multimode fiber
MNDS	Multinetwork Design System
modem	modulator/demodulator
MOP	Management Operations Protocol
MOV	metal oxide varistor
MPDU	message protocol data unit
MSAP	MAN service access point
MSAT	mobile satellite
MSAU	Multistation Access Unit
MS-DOS	Microsoft Disk Operating System
MSP	Message Service Protocol
MUX	multiplexer
nak	negative acknowledgment
NAT	Network Address Translation
NAU	Network Access Unit
NBP	Name Binding Protocol
NCP	NetWare Core Protocol
NetBEUI	NetBIOS Extended User Interface
NetBIOS	Network Basic Input/Output System
NetID	network identification
NIC	network interface card
NISDN	Narrowband ISDN
NIU	Network Interface Unit
NLP	NetWare Lite Protocol
NLSP	NetWare Link Service Protocol
NMP	Network Management Protocol
NNTP	Network News Transfer Protocol
NOS	network operating system

Appendix B

Acronym or Abbreviation	What It Stands For
NPAI	Network Protocol Address Information
NPDU	network protocol data unit
NSP	Network Services Protocol
NT	New Technology
NTFS	New Technology File System
NTP	Network Time Protocol
NTU	network terminating unit
NVP	Network Voice Protocol
NVRAM	non-volatile RAM
ODBC	Open Database Connectivity
OS	operating system
OSI	Open Systems Interconnection
OSPF	Open Shortest Path First
PAD	Packet Assembler/Disassembler
PAP	Packet Access Protocol *or* Password Authentication Protocol
PC	personal computer
PCN	personal communications network
PCSN	private circuit-switching network
PD	public domain
PDA	Personal Digital Assistant
PING	Packet Internet Groper
PIR	Protocol-Independent Routing
PMX	Private Message Exchange
POP	Point of Presence
POST	Power On Self Test
POT	point of termination
POTS	plain old telephone service
PPP	Point-to-Point Protocol
PPSDN	public packet-switched data network
PPSN	public packet-switched network

Acronyms and Abbreviations

Acronym or Abbreviation	What It Stands For
PRI	Primary Rate Interface
PRMD	private mail domain
PSAP	presentation service access point
PSDN	packet-switched data network or public switched data network
PSI	Packet Switching Interface
PSN	packet switched network
PSTN	public switched telephone network
PTN	public telephone network
PVN	private virtual network
PVT	permanent virtual terminal
QoS	Quality of Service
RAID	Redundant Array of Inexpensive Disks
RAM	random access memory
RARP	Reverse Address Resolution Protocol
RAS	remote access server
RD	routing domain
RDP	Remote Data Protocol
RF	radio frequency
RFC	Request for Comments
RFI	radio frequency interference
RFP	Request for Proposal
RIF	Routing Information Field
RIP	Router (Routing) Information Protocol
RJ	registered jack
RMHS	Remote Message Handling Service
RMON	Remote Monitoring
ROM	read-only memory
RSO	regional standards organization
RSPX	Remote Sequenced Packet eXchange
RTF	Rich Text Format

Appendix B

Acronym or Abbreviation	What It Stands For
RTMP	Routing Table Maintenance Protocol
RUP	Routing Update Protocol
SAGE	Semiautomatic Ground Environment
SAP	Service Advertisment Protocol
SAS	Single Attachment Station
SDDP	Short Datagram Delivery Protocol
SDLC	Synchronous Data Link Control
SDM	space-division multiplexing
SDRP	Source Demand Routing Protocol
SFS	Shared File System
SHTTP	Secured HyperText Transport Protocol
SIMP	Satellite Information Message Protocol
SLIP	Serial Line Interface Protocol
SMP	Session Management Protocol *or* symmetric multiprocessing
SNA	Systems Network Architecture
SNAP	Subnetwork Access Protocol
SNMP	Simple Network Management Protocol
SNMPv2	Simple Network Management Protocol version 2
SPARC	Standards Planning and Review Committee *or* Scalable Processor Architecture
SPF	Shortest Path First
SPP	Sequenced Packet Protocol
SPX	Sequenced Packet eXchange
SQL	Structured Query Language
SRB	source route bridging
STM	Synchronous Transfer Mode
STP	shielded twisted pair
SVC	switched virtual circuit
SWAN	satellite wide area network
Tb	terabit
TB	terabyte

Acronyms and Abbreviations

Acronym or Abbreviation	What It Stands For
Tbps	terabits per second
TBps	terabytes per second
TCP	Transmission Control Protocol
TCP/IP	Transmission Control Protocol/Internet Protocol
TDM	time-division multiplexing
TDR	Time Domain Reflectometry
telco	telephone company
TFTP	Trivial File Transfer Protocol
THz	terahertz
TLAP	TokenTalk Link Access Protocol
TP	Transport Protocol
TPAD	Terminal Packet Assembler/Disassembler
TR	Token Ring
TSS	Time-Sharing System
TTL	time to live
TTP	Timed Token Protocol
UDP	User Datagram Protocol
UHF	ultrahigh frequency
ULP	upper-layer protocol
URL	Uniform Resource Locator
USB	Universal Serial Bus
UTP	unshielded twisted pair
UTTP	unshielded telephone twisted pair
VARP	VINES Address Resolution Protocol
VHF	very high frequency
VINES	Virtual Integrated Network Services
VIP	VINES Internet Protocol
VLAN	virtual local area network
VPN	virtual private network
VTP	Virtual Terminal Protocol
WAN	wide area network

Appendix B

Acronym or Abbreviation	What It Stands For
WDM	wavelength-division multiplexing
WEP	Wireless Equivalent Privacy
WINS	Windows Internet Naming Service
WLAN	Wireless Local Area Network
WPAN	Wireless Personal Area Network
WWW	World Wide Web
XNA	Xerox Network Architecture
XNS	Xerox Network System
ZIP	Zone Information Protocol

Appendix C

Glossary

10BaseT
A type of network that specifies the use of a baseband transmission on an 802.3 Ethernet network using unshielded twisted pair and able to transmit up to 10 Megabits per second (Mbps).

abstract syntax
The format of data in the Application layer of the OSI model before it is converted to any other format by the Presentation layer.

access layer
The layer in which network devices act as an intermediary connection point between the end systems and the distribution layer.

access point (AP)
A device or software used to extend a wired LAN by transmitting wireless data signals to wireless nodes.

access rights
Security properties assigned to a network object or resource that define the level of access for a user. Common levels of access rights include read, write, change, and no access. Also known as *access permissions*.

ack (acknowledgment)
The message used to respond positively to a request for synchronization when establishing a communication session between two devices.

adapter
A general term for a network interface card.

ad-hoc mode
A manner of transmitting wireless signals directly between devices or stations without the use of a wireless access point. Also called *peer-to-peer mode*.

agent
Software that collects MIB information about a managed device, and provides the information to the management console.

ALOHAnet
A network that connected the Hawaiian Islands using radio transmissions.

alternating current (AC)
An electrical current that reverses its direction at regularly recurring intervals.

American National Standards Institute (ANSI)
An organization that coordinates the development of standards, including those in the area of communications and networking. It also approves U.S. national standards, and is a member of the IEC and the ISO.

ampere
The unit of measurement of electrical current. Also called an *amp*.

amplitude
The maximum value of a signal as measured from its average state, which is usually zero.

amplitude modulation (AM)
Signals in the same frequency that have different amplitudes.

analog
A type of signal that uses a continuous waveform combining amplitude, frequency, and phase.

analog signal
The fluctuation of an electrical signal or other source that has the characteristic of being continuous rather than having discrete points, as is the case with digital signals.

AP
See *access point*.

AppleShare
The network operating system developed by Apple Computer that runs on a Macintosh server.

AppleTalk
Apple's proprietary protocols that are built into the Mac OS and allow Apple Macintoshes to communicate right out of the box.

Glossary

Application layer
The top layer of the OSI model. The Application layer formats data for a particular function such as network printing, electronic mail, or web viewing.

archive bit
A bit setting for a file that is used by backup programs to determine when the file was last backed up. The bit is reset after every differential and normal backup.

ARCNET
A network access method that uses tokens such as Token Ring, but is much less expensive.

ARP table
A list of MAC addresses mapped to the corresponding IP addresses of the workstations that they represent. The ARP table is maintained in the memory of the NIC.

ARPAnet
The predecessor to the Internet, ARPAnet was developed by the Department of Defense's Advanced Research Projects Agency to provide reliable communication, even in the event of a partial network failure.

asynchronous
A transmission type that uses start and stop bits for bit synchronization.

Asynchronous Transfer Mode (ATM)
The transmission of digital signals using different frequencies and timing, allowing for multiple transmissions to occur through the use of packet switching.

ATM switch
A network device used by telecommunications companies like the local telephone company to support multiple connections on an ATM network.

attenuation
The degradation of a signal because energy is lost as it moves down the wire.

authentication
The process of a server validating the username and password submitted by a user during login. If the authentication is successful, the server returns the user's security rights; if unsuccessful, the user is denied access.

autoloader
A multidrive tape device that holds several tapes and automatically switches to the next tape when a tape is filled.

B channel
Bearer channel. A 64Kbps channel used in ISDN that provides full-duplex communication for the user.

backbone
The main connection point for multiple networks that carries the bulk of the traffic between different networks or multiple network segments, typically high speed.

backoff
After a collision, all devices on an Ethernet network wait a random amount of time before transmitting data.

bandwidth
The total capacity of a connection or transmission measured in bits.

baseband
Transmissions that occur using a single fixed frequency.

baselining
Creating a level of reference to use in a performance analysis.

binary
A number system that uses base 2 rather than the decimal numbering system that uses base 10. In a binary system, information is represented as either a 1 or 0.

BIOS (basic input/output system)
The BIOS service, located on a computer's ROM chip, enables the hardware and software to communicate with each other.

bit
The smallest form of information in a computer that has a value of either 1 or 0.

bit stream
Describes the transmission of a contiguous group of bits in a digital signal.

bit synchronization
A function that is required to determine when the beginning and end of the transmission of data occurs.

blackout
A complete loss of electrical signal.

Bluetooth
A wireless technology developed jointly by companies such as Ericsson, IBM, Intel, and Nokia to support the exchange of data between wireless devices.

BNC connector
Connector used with coaxial cable.

bottleneck
A device on the network that is slow in transmitting data, causing delays in the delivery of data.

bridge
A device that will filter or forward data packets between two or more networks or network segments based on the MAC address of the destination. It works between networks or segments that use the same protocol.

Broadband ISDN (BISDN)
An extension of ISDN that is capable of supporting voice, data, and video at very high speeds using ATM as the transmission system.

broadcast
Data that is sent out to all devices on the network.

broadcast domain
All devices on a network that can receive the same broadcast packet.

BSD UNIX
A version of UNIX developed at the University of California, Berkeley. BSD UNIX is available for free from the Berkeley Software Distribution user group.

bus topology
A physical topology that utilizes a single main cable to which devices are attached.

cancellation
The process in which two wires are twisted together to prevent outside interference or cross talk.

Carrier Sense Multiple Access/Collision Detection (CSMA/CD)
The access method that is the basis for Ethernet and works at the MAC sublayer of the OSI Data Link layer.

cell
A fixed-size packet. In ATM networks, the cell size is fixed at 53 bytes.

cell switching
A form of switching, similar to packet switching, that uses fixed 53-byte cells to transport data.

central office (CO)
The local telephone company office in which all local loops connect and provide switching of telephone circuits.

central processing unit (CPU)
A microprocessor chip, the CPU is the brain of the computer; it uses mathematical functions to perform calculations and complete tasks.

chassis
A metal case with slots that will accept different electronic boards that can be hubs, routers, or bridges.

checksum
A way of providing error checking by calculating a value for a data packet that can be used by the destination device, which recalculates the value of the packet and can determine if the packet was received intact.

chips
The chunks of data that are transmitted within a direct-sequence modulation transmission.

Glossary

circuit
The path for the usage of electricity.

cladding
A reflective material that surrounds the fiber-optic strand for the purpose of focusing the light down the fiber-optic strand and preventing the loss of light.

client-server network
A network architecture that combines the processing of both the workstation and the server to perform a task.

coaxial cable
A type of media that consists of a single copper wire surrounded by insulation and a metal shield of foil or braid, and covered with a plastic jacket made of PVC or plenum.

collision domain
An area of the network, bound by bridges, switches, or routers, in which collisions are propagated.

common mode
Electrical power problems between the hot or neutral wires and the ground wires in an electrical outlet.

communications protocol
A set of rules that defines how communications will occur.

Compatible Time-Sharing System (CTSS)
Developed in 1961 at MIT, CTSS had a capacity of up to 30 modems to give terminals access to run tasks concurrently. CTSS was the precursor to operating systems such as UNIX.

conduit
Metal or plastic pipe or raceway used to protect the cable runs from physical damage or electrical interference.

connectionless
When communication occurs between devices, there is no acknowledgment that data has been received.

connection-oriented
During communication between two devices, the receiving device will acknowledge to the sender that it has received the data. If part of the data is not received, the sender retransmits the data.

continuity test
A pass-or-fail test used to determine if an electrical circuit exists between a pair of wires.

cross talk
The electromagnetic interference that occurs when the electrical signal on one wire changes the electrical properties of a signal on an adjacent wire.

crossover cable
A special cable that reverses the transmit and receive wires from one end to the other to directly interconnect devices without a hub.

D channel
Data channel. A 16Kbps (BRI) or 64Kbps (PRI) full-duplex ISDN channel.

daisy chain
Using a cable to connect devices together from one to the next so that the output of one device is connected to the input of another to form a chain.

data
A term used to represent the information that is transferred and formatted in the top three layers of the OSI model.

Data center
A facility that supports large numbers of computers and other technology.

data compression
The reformatting of data to make it smaller.

Data Link layer
Layer 2 of the OSI model, the Data Link layer receives data from the Network layer, and packages it as frames to be sent onto the network by the Physical layer.

datagram
A data packet that has had the network address added to it as part of the encapsulation process.

decimal
A numbering system that uses base 10.

deencapsulated
Layer information is stripped from a packet as it moves up the OSI layers.

default gateway
A device, usually a router, that connects two or more different networks.

demarc (demarcation)
The location in your building or office where the local phone company's responsibility for the lines and equipment ends.

demultiplex
Separating the bits of information from a multiplexed frame back into the individual voice or data signals.

destination
A device on a network that is the recipient of data transmitted by the sender.

devices
Hardware that is capable of attaching to a network and communicating with other hardware. Examples are computers, printers, routers, repeaters, hubs, and bridges.

dialog control
A function of the Session layer that determines which device will communicate first and the amount of data that will be sent.

dialog separation
The use of markers within the data to determine whether all information received is intact.

dial-up
The process of connecting to the Internet using a modem and telephone line.

dielectric
A type of insulation material.

digital
A type of signal that uses pulses to send binary signals across media. One pulse represents a 1; a lack of a pulse represents a 0.

Digital Linear Tape (DLT)
Developed by DEC as a reliable, high-speed backup solution for mainframes, DLT technology has been considered one of the most stable and widely proven tape backup technologies.

digital signal
The transmission of data using a discontinuous source, which can be accomplished with electricity or electromagnetic fields.

direct current (DC)
An electrical current that flows in one direction only and is substantially constant in value.

direct-sequence modulation
A type of spread-spectrum signal that transmits chips that can be reformatted by the receiver at different frequencies.

disk mirroring
A disk-redundancy technique that uses a primary disk on which data is stored and a secondary disk of equal capacity to which an exact duplicate of the primary disk data is copied.

disk striping with parity
A hard disk fault-tolerance technique that uses a minimum of three hard disks of the same size. Data is written across all three hard disks, along with some error-correction information called *parity*. If one hard disk fails, a new disk can be rebuilt using the information on the other disks.

distribution layer
The devices that provide access to the network backbone or to the core layer.

DoD model
The term used to describe the conceptual model for the TCP/IP protocol suite.

Glossary

domain
In Microsoft terminology, *domain* refers to the domain model for organizing NT servers within a company.

downtime
The amount of time that a service, such as e-mail or a file server, is not functioning.

driver
Software used to interface between a device and the operating system.

drop cable
A portion of cable that attaches from an end device to a vampire tap attached to a Thicknet cable.

DS0
A digital channel used for data or voice transmissions.

dynamic routing
Internetwork routing that adjusts automatically to network topology or traffic changes based on information it receives from other routers.

earth ground
A wire that allows excess electricity to go into the ground. It has a potential electrical charge, which measures zero. Also called a *ground*.

e-commerce
Short for *electronic commerce*. The buying and selling of goods or services over the Internet.

EIA/TIA
Electronics Industries Alliance/Telecommunications Industry Association.

electron
A negatively charged particle.

Electronics Industries Alliance (EIA)
A group of professionals that develops electrical transmission standards. The EIA and TIA work jointly to develop many communications standards.

electrostatic discharge
The displacement of static electric buildup.

EMI (electromagnetic interference)
When the electric field created by electricity on one wire changes the form of an electrical charge on an adjacent wire.

encapsulated
The process of enclosing a packet of data with information from the current layer as it passes down the OSI layers.

encapsulation
In LAN protocols, the process of adding network address information to a data packet. When the data packet has been encapsulated, it is called a *datagram*.

encryption
The process of converting data into a random set of characters that is unrecognizable to everyone except the intended recipient.

end system
A device participating in a communication with another device.

Ethernet
A network access technology that incorporates Carrier Sense Multiple Access/Collision Detection (CSMA/CD) to access the media and uses a logical bus topology.

extended star topology
A physical topology that connects additional hubs to a central star topology to add additional devices.

FAQ (frequently asked questions)
A list of specific questions about the most common problems encountered by users, followed by the answers.

Fast Ethernet
A network technology defined by the IEEE 802.3u standard for 100Mbps networks.

fault
A memory error that occurs when the CPU is unable to find the location of data stored in the swap file.

fault tolerance
The ability of a device or system to continue to operate despite software problems, power or hardware failures.

Federal Communications Commission (FCC)
A government agency that reports directly to the U.S. Congress. The FCC is charged with regulating interstate and international communications by radio, television, wire, satellite, and cable.

feedback loop
Assessing an event and using that information to improve the process before the event is repeated.

Fiber Distributed Data Interface (FDDI)
A type of network that uses 100Mbps dual-ring token passing with fiber-optic cable.

fiber-optic cable
A type of media that uses very thin glass or plastic to transmit light signals.

File Transfer Protocol (FTP)
A protocol that defines how files are transferred to and from an FTP server and a client computer.

firewall
Devices used with specialized software to secure a LAN from access by unauthorized sources

flow control
A Transport layer feature that manages the flow of data. If the receiving device is unable to process incoming data, it sends a message to halt the transmission; when able to continue, it tells the sender to begin transmitting again.

Frame Relay
A WAN connection protocol that provides multiple virtual circuits. It is more efficient and less expensive than many WAN connections, and is currently being considered as the replacement for the X.25 protocol.

frequency
The number of cycles completed in one second.

frequency hopping
A type of spread-spectrum signal that switches the transmission of data between frequencies based on a timing pattern that controls when the switch should occur.

frequency modulation (FM)
Signals that may have the same amplitude, but have different frequencies.

full duplex (FDX)
On a computer network, data can travel in both directions simultaneously.

gateway
A device that performs the translation of information from one protocol stack to another.

gauge
The measurement of the diameter of electrical wire.

generator
A machine that converts mechanical energy into electrical energy.

Get Nearest Server (GNS)
A process that uses SAP broadcasts to request services from servers available on the network.

GHz
Gigahertz; one billion hertz.

gigabit
A unit of measurement defined as 1,073,741,824 (approximately one billion) bits.

Gigabit Ethernet
A network technology defined by the IEEE 802.3z standard for 1,000Mbps or 1Gbps networks.

gigabit Point of Presence (gigaPOP)
A site that is considered a main backbone provider for Internet 2, and is capable of supporting internetwork speeds in the gigabit range.

half duplex (HDX)
On a computer network, data travels in either direction, but not at the same time.

Glossary

hard disk
A large-capacity magnetic storage device with read and write capabilities.

hardware address
The term used to describe the MAC address on a Macintosh computer.

hertz (Hz)
A unit of measurement for the frequency of cycles in a signal.

hexadecimal
A numerical system that uses the first six letters of the alphabet to extend the possible digits to 16 beyond the 10 available in the decimal system.

hot wire
The electrical wire in an outlet that is the source of the incoming electrical signal from the power company.

HTML (HyperText Markup Language)
A scripting language, developed in the early 1990s, which uses a browser for viewing documents on the Internet.

hub
A network connectivity device that connects multiple network nodes together. Used primarily with Ethernet, it forwards all traffic it receives from one port to all other ports.

hybrid architecture
On a local area network, the use of both peer-to-peer and client-server network architectures.

hybrid topology
A physical topology that combines one or more physical topologies, including the bus, ring, and star.

HyperText Transfer Protocol (HTTP)
A protocol that defines how web pages are processed by web servers.

IBM data connector
The type of connector used in IBM networks with shielded twisted-pair cabling.

impedance
The opposition to electrical current, measured in ohms.

infrared
Electromagnetic waves that can be used to transmit and receive wireless data transmissions.

infrastructure mode
A manner of transmitting wireless signals that utilizes an access point as an extension of a wired network. This allows communication between wireless nodes or the wired network.

Institute of Electrical and Electronics Engineers (IEEE)
An organization that develops and adopts communications and networking standards, including the 802 series of LAN standards.

Integrated Services Digital Network (ISDN)
A WAN communications protocol offered by telephone companies that allows telephone lines to carry data, voice, and other types of transmissions.

interference
Also referred to as *electromagnetic interference (EMI)* or *radio frequency interference (RFI)*, interference is a disturbance in the electromagnetic field of a circuit that may make the signal distorted or unrecognizable.

intermediate distribution facility (IDF)
A wiring closet that is an extension of the MDF; used to reach devices that are too far from the MDF.

intermediate system
Any device that is used to assist in transporting data between two communicating end systems.

International Organization for Standardization (ISO)
An international organization that develops standards, including those relevant to networking. The ISO developed the OSI model used as the theoretical model for intercommunication between networking devices.

International Telecommunications Union (ITU)
An international organization that develops communication standards.

Internet 2
The second Internet, connecting major university campuses, research institutes, and government agencies across the country for research and collaboration using Gigabit speed connections.

Internet service provider (ISP)
A company that offers (to businesses and the public) access to the Internet by selling services such as dial-up accounts that use modems.

internetwork
Two or more connected networks of similar or different communication types.

InterNIC
Formerly, the organization responsible for the managing the allocation of IP addresses and domain names for all organizations.

interoperability
The capability of hardware and software to work together to complete a task, regardless of the manufacturer. Standards are often created to dictate the way manufacturers are supposed to develop products using a specific technology.

inventory
A spreadsheet or database listing details about the equipment owned or leased by a company.

IPX/SPX
Novell's proprietary protocol suite that provides end-to-end communication on Novell NetWare networks.

ISA (Industry Standard Architecture)
A motherboard technology that is used to connect expansion cards (NICs, sound cards). Much slower than PCI, ISA was originally designed for the IBM PC and was used in the AT, 386, and 486 models.

joule
A unit of energy equal to the work done by a force of one newton acting through a distance of one meter.

keep-alive message
A data packet sent between Session layers to keep inactivity from causing the connection to close down.

laser
An extremely focused beam of light. The term comes from "light amplification by simulated emission of radiation."

latency
A delay in the transmission of data usually caused by excessive traffic.

leased line
A type of dedicated transmission connection for the use of a private customer typically leased from a telecommunications company.

legacy
From the past.

light-emitting diode (LED)
Electrical signals converted into light by a semiconductor.

line conditioner
A device that helps to regulate electrical power and provide protection to electronic devices. Also known as a line conditioner.

Linux
Linux is a variant of UNIX that was originally created by Linus Torvalds at the University of Helsinki, but now includes some of BSD UNIX and SVR4.

local area network (LAN)
The interconnection of computers, hubs, and other network devices in a limited area, such as a building.

local exchange carrier (LEC)
The local telephone company and the CO switches that provide local service.

Glossary

local loop
The two-wire copper telephone cable that runs from a residential or commercial location to the central office of the telephone company.

logical bus topology
A system in which data travels in a linear fashion away from the source to all destinations.

Logical Link Control (LLC)
The LLC sublayer establishes whether communication is going to be connectionless or connection-oriented at the Data Link layer.

logical ring topology
A system in which data travels in a logistical ring from one device to another.

logical topology
The way in which data is transmitted throughout a network.

MAC address
The address of the device that is found on the NIC. Also known as the *physical address*.

MAC broadcast
A broadcast packet sent on the network with the MAC address of FF-FF-FF-FF-FF-FF, which requires all devices to pass the packet to the Network layer for address identification.

Macintosh
A type of computer that Apple Computer introduced in 1984 that was distinguished by its easy-to-use graphical user interface (now called the Mac OS).

main distribution facility (MDF)
The central wiring closet that is used to house the core network devices such as the core hub, switch, and routers. All other wiring closets connect directly to the MDF.

mainframe
A large, powerful computing machine that stores and processes information, and that runs applications for the terminals that are attached to it.

management information base (MIB)
A database that contains the definitions and values of objects for a managed device.

MAU (Multistation Access Unit)
The central hub in a Token Ring network that is wired as a physical star.

media
The physical transmission pathway that connects devices in a network—cable, for example.

Media Access Control (MAC)
The MAC sublayer adds the destination and source physical addresses to a frame before sending it on to the Physical layer.

mesh topology
A physical topology in which all devices are interconnected. Also called a *net topology*.

metal oxide varistor (MOV)
An electrical component of a surge suppressor or line conditioner that absorbs excess electrical signals.

metropolitan area network (MAN)
Two or more LANs interconnected over high-speed connections across a city or metropolitan area.

metropolitan exchange
The main telecommunication switches found in a city to centrally connect all COs.

MIC (memory in cassette)
A flash memory chip imbedded in an AIT cassette that stores the catalog of the tape's contents.

micron
A measurement equal to 1/1,000,000 of a meter or 1/25,000 of an inch.

microsecond
One-millionth of a second.

mnemonic
A way of remembering information by using a phrase, song, or some other method.

modem
A device that turns digital signals to analog, and vice versa, for communication on regular telephone lines.

motherboard
The main board in a computer that manages the communication and function of all information.

multichannel trunk
A four-wire cable, used for dedicated leased lines, which is capable of supporting up to 24 voice conversations or 1.536Mbps of data.

multimode
Fiber-optic cable that supports multiple transmission signals using LED or laser light.

multiplex
Sending multiple voice or data signals across a single wire by inserting bits of information from each of the signals into a frame.

multiplexing
A process that allows multiple signals to transmit simultaneously across a single physical channel.

multiprocessing
Multiple processors within a device working on the same task.

multitasking
The processor's capability to perform several tasks simultaneously.

near-end cross talk
Interference on a wire from the electromagnetic field of another wire that was terminated incorrectly.

needs assessment
A process of collecting information about the needs of individuals and an organization to improve the performance of a system.

NetBEUI (NetBIOS Extended User Interface)
Pronounced "net-booee," NetBEUI is a Microsoft protocol that was originally designed for Microsoft LAN Server software. NetBEUI is considered a fast protocol, but it cannot be routed between networks.

NetWare
A network operating system created by Novell.

network address
The address assigned by the network administrator, based on the network where the device is located. Also known as the *logical address*.

network administrator
A person whose job it is to manage and support the network infrastructure, including supporting the needs of users.

network architecture
A design that reflects the intended use of hardware and software that will allow computing devices to communicate with each other.

network interface card (NIC)
The internal hardware installed in computers and other devices that allows them to communicate on a network.

Network layer
Layer 3 of the OSI model, the Network layer takes the packets passed down from the Transport layer, and adds the appropriate network addresses to them.

network management system (NMS)
The use of hardware and software by a network administrator to monitor and manage the network.

network operating system (NOS)
Software designed specifically for use on a network server to provide multiple services and a high level of performance.

network operations center (NOC)
The central location in a network where all management is maintained, and core devices such as routers are located.

Glossary

network segment

The area of the network, bound by bridges or switches, in which collisions are propagated; or the area bound by a router to prevent the propagation of broadcasts.

neutral wire

The electrical wire in an outlet that provides the return path for the electrical signal, which completes the circuit. It has a ground potential of zero.

node

Any device connected to the network that is capable of sending and receiving data.

noise

Another term for *interference*.

normal mode

Electrical power problems between the hot and neutral wires in an electrical circuit or outlet.

Novell Directory Services (NDS)

A set of software services created by Novell to provide access to a directory of information about network entities. It allows an administrator to organize and manage users, servers, and other devices on the network.

NSFNET

The name for the network backbone funded and built by the National Science Foundation that connected many isolated networks to the ARPAnet.

OC (optical carrier) levels

Line speeds in SONET networks.

ohm

A measurement that translates to one unit of resistance.

orthogonal frequency division multiplexing (OFDM)

A type of radio signal used in wireless networks that divides the data signal across 48 separate sub-carriers. This leads to high data transmissions and superior performance.

OSI (Open Systems Interconnection) model

The communications model developed and adopted by the ISO in 1977. It defines how hardware and software should be developed to support specific functions for communication between devices.

packet

Describes the form of data after it has passed through the Transport layer. A packet is a group of bits that includes a header, data, and trailer; and can be transmitted over a network.

password

A combination of letters, numbers, and symbols either assigned by the administrator or selected by the user that is used in conjunction with the username to gain security access to a resource.

patch cable

As specified by the EIA/TIA 568 standard, a network cable that is used to connect an end system to a cable that is terminated in the wall.

patch panel

Also referred to as the *horizontal cross-connect*, the patch panel is mounted on a rack and is used to connect the cables from the wall outlets to the hubs in the wiring closets.

PCI (Peripheral Component Interconnect)

A motherboard technology developed by Intel that improves the speed of communication between an expansion device and the CPU to take advantage of faster CPU speeds.

PDA

Personal Digital Assistant. A handheld device that can provide the functions of a personal organizer, phone, and fax sender, as well as web browsing and other network features.

peer-to-peer mode

A manner of transmitting wireless signals between devices or stations without the use of an *access point*. See *ad-hoc mode*.

peer-to-peer network
A network architecture that allows 10 or fewer users to effectively share files and folders on their computers with other users on their networks.

Physical layer
The bottom layer of the OSI model, the Physical layer specifies the type of media to be used, the transmission format, and the topology of the network.

physical topology
The physical layout of a network's media.

Ping
An application used to test whether a connection exists between two remote devices on the Internet. Requires the TCP/IP protocol.

plenum
A type of plastic that meets fire safety standards. It does not burn or create toxic fumes when exposed to heat or fire.

point of entry (POE)
The location in the building where the telephone lines enter from the street.

polling
A process used by the SNMP protocol to periodically check for updated management information from the devices.

port
The female interface on an internetworking device. Used with jacks to form a connection.

power conditioner
A device that helps to regulate electrical power and provide protection to electronic devices. Also known as a line conditioner.

pre-emptive multitasking
The capability of some NOS software, such as Windows NT and UNIX, to manage tasks so each application gets equal access to the processor.

premise
The location of equipment and wiring at the end-user site.

Presentation layer
Layer 6 of the OSI model. The Presentation layer manages the conversion from data structures used by the computer to a form necessary for communication over the network.

print queue
A location on a server or workstation that stores print jobs that are waiting for a print device.

processor
Another term for CPU.

promiscuous mode
The collection of data from the network without other devices having any knowledge of the activity.

proprietary
The sole property of one individual or organization.

protocol
A set of rules used to define communication between two devices.

public data network (PDN)
A fee-based network whose main concern is to provide computer communications to the public. It may be operated by a private company or a government.

public switched telephone network (PSTN)
The telephone infrastructure that relies on circuit switching or other switching technology to open and close circuits for voice conversations. Also known as *plain old telephone service (POTS)*.

punch block
Similar to a patch panel, except that it mounts directly on plywood on the wall in the wiring closet. Punch blocks are typical of telephone cable installations.

Glossary

PVC
The plastic outer housing typical of data cable.

redundancy
A method of providing fault tolerance by duplicating a service or hardware.

repeater
A network device that regenerates and propagates signals between two network segments.

Request for Comment (RFC)
A public request for feedback about a standard being developed.

RFI (radio frequency interference)
Interference with transmissions over copper wires caused by radio signals.

ring topology
A physical topology in which all devices are connected in a circle, providing equal access to the network media.

RJ-45
The type of connector used with twisted-pair cabling.

routed protocol
A LAN protocol that can be transmitted to other networks by way of a router; examples include AppleTalk, DECnet, IP, and IPX.

router
A device used to select the best path for data travel to reach a destination on a different network.

routing
The use of a routing protocol to select the best path to the appropriate network.

routing protocol
A protocol that uses a specific algorithm to route data across a network.

routing table
A table that keeps track of the routes to networks and the costs associated with those routes.

sag
A reduction in electrical signal to 80 percent of the normal voltage. Also known as a *brownout*.

satellite
In telecommunications, a device that is sent into earth's orbit to travel around the earth and provide telecommunications services for voice and data.

SC connector
Fiber-optic connector that is square in form and can be keyed.

SCSI (Small Computer System Interface)
A high-speed interface for connecting hard disks, CD-ROMs, and other peripherals to a computer.

security
In data processing, the ability to protect data from unauthorized access, theft, or damage.

serial port
A physical interface on a computer that is commonly used to attach a mouse, printer, or modem. A common type of serial interface is RS-232.

server
A computer used to centralize resources, administration, and security for a network and its users by using specialized hardware and software.

service
An application on a server that provides greater functionality for the end user.

session
A communication channel that is created and maintained between two networked devices in order to transfer data.

Session layer
Layer 5 of the OSI model, the Session layer creates, maintains, and terminates communication between devices on a network.

share-level security
The configuration of a shared folder with access permissions using only a password.

shielded twisted pair (STP)
A type of media that consists of pairs of twisted wires that are insulated by foil and covered in a plastic jacket.

Simple Mail Transfer Protocol (SMTP)
A protocol that specifies how electronic mail is to be delivered to its destination.

simplex
A transmission flow that allows communication in only one direction.

sine wave
A waveform that represents the positive and negative changes in an AC electrical current.

single-mode
Fiber-optic cable that has only one cable and allows a single transmission signal at one time.

SMA connector
Fiber-optic connector that has a screw-on connector.

source
The device in which data being sent over a network originates.

Spanning Tree Protocol (STP)
A layer 2 protocol used in bridges and switches to identify the shortest path to any device on a local area network.

spike
An increase in the electrical signal of at least twice the normal voltage for 10 to 100 microseconds.

square wave
The waveform used to represent digital bits in their electrical form.

ST connector
Fiber-optic connector that uses a twist-and-lock mechanism.

stack
Several hubs from the same vendor that are stacked together as one larger hub.

standard
A set of rules or procedures that is officially accepted.

star topology
A physical topology that connects networking devices to a central hub.

star-bus
A mixed physical topology that includes a star and a bus, and that is connected together using a device, such as a hub.

static electricity
The electrical energy produced by friction.

static routing
A type of routing in which the network administrator manually configures a route into the router's routing table.

subnet mask
A special 32-bit address used to indicate the bits of an IP address that are being used to create subnets. Sometimes called a *subnet address* when shown as a decimal dotted notation number.

subnetting
A method of splitting an IP network address into smaller groups of IP addresses that can be used on different networks.

subnetwork (subnet)
A smaller segment of a larger network that is created by a network administrator through the use of a subnet mask.

Subnetwork Access Protocol (SNAP)
An Internet protocol that specifies the encapsulation method for IP datagrams and ARP messages, and provides a link between a subnetwork device and an end system.

subscriber
A home or office that utilizes the public telephone network.

Glossary

SuperDLT
The modernization of the DLT format taking advantage of new optical and magnetic technologies that can place more tracks of data on the tape media. See also *Digital Linear Tape (DLT)*.

surge
An increase in the electrical signal to 110 percent or more of the normal voltage for less than 2.5 seconds.

surge suppressor
A device that is used to absorb excess electrical signals to prevent damage to electrical devices.

swap file
A part of the hard disk that is used as a temporary storage space when the computer has used all available RAM.

switch
A multiport switch that provides LAN connectivity for multiple network nodes. The device filters or forwards data packets to the correct port for a device based on the MAC address of the destination, thus eliminating unnecessary network on other ports.

switched circuits
Once a mechanical feature, switched circuits are used to create logical connections between two points on a switch in a CO.

syn (synchronization)
The message used to establish communication between two or more systems.

synchronous
A transmission type that uses timing for bit synchronization.

Synchronous Optical Network (SONET)
A high-speed synchronous network that runs on optical fiber at speeds of up to 9.9Gbps.

T-carrier
A technology developed by AT&T that was designed to provide multiple voice conversations over a single four-wire cable called the *trunk*.

TCP/IP protocol suite
Represents several different protocols that may run in tandem with, independently of, or in place of Transmission Control Protocol (TCP) and Internet Protocol (IP).

Telecommunications Industry Association (TIA)
An organization that works to develop and adopt telecommunications technologies standards. The TIA and EIA work jointly to develop many communications standards.

terminal
An input/output device that relies solely on a minicomputer or mainframe for processing, storing, and viewing data.

termination
The location where a cable ends and connects to a jack or patch panel.

terminator
A device used to terminate the ends of the main cable in networks implementing a physical bus topology and using Thinnet cabling.

Thicknet
A type of coaxial cable, typically used in older networks, that is approximately 0.4 inches in diameter. Also called *10Base5*.

Thinnet
A type of coaxial cable, typically used in older networks, that is approximately 0.2 inches in diameter. Also called *10Base2*.

threshold
A value set by the network administrator that, when exceeded, indicates there is a critical situation.

Time Domain Reflectometry (TDR)
A technology found in cable testers, TDR sends a signal down a copper wire and waits for the signal to bounce back. Based on the amount of time it takes for the signal to return, the TDR cable tester can determine the length of the cable.

Appendix C

timing
A function used in synchronous transmissions, in which two systems that are communicating each agree on the time that it takes to transmit one bit of data.

token
A frame passed on a Token Ring network. Possession of the frame allows a device to transmit data on the network.

Token Ring
A network access method that incorporates token passing to access the media. Data travels in only one direction in a Token Ring environment.

Trace Route
A utility that is used to identify the path a packet has taken to reach a destination device on the Internet.

transfer syntax
The format of data after it has been converted by the Presentation layer into a "common language" format, typically ASCII.

transformer
A device that uses induction to convert a primary circuit into a different voltage in a secondary circuit.

transponder
A device for receiving radio signals that then converts the signals and transmits them in a different format.

Transport layer
Layer 4 of the OSI model, the Transport layer takes the data from the Session layer and breaks it up into segments that can easily be transmitted by the lower-layer hardware.

Ultrium (LTO)
Ultrium is the implementation of a reliable, fast, and large-capacity open tape technology called Linear Tape-Open (LTO) ,which was jointly developed by IBM, HP, and Seagate.

Underwriters Laboratories (UL)
UL is responsible for testing the safety of electrical wires including the types of cable used in networks.

uninterruptible power supply (UPS)
A device that converts stored direct current to alternating current to provide power in the event of a reduction or loss in the primary electrical signal.

UNIX
An operating system that supports multiple users, multitasking, and (in many cases) multiple processors. UNIX was created by AT&T in 1969. Today, there are several variations of UNIX.

unshielded twisted pair (UTP)
A type of media that consists of one or two pairs of twisted wires and is covered in a plastic jacket.

uplink port
The port on a hub or switch that is used to extend the network to another hub, repeater, or switch.

vampire tap
The type of connector used to attach a drop cable to a Thicknet backbone cable.

very high performance Backbone Network Service (vBNS)
The gigabit network developed and managed by MCI in cooperation with the National Science Foundation and other agencies.

virtual circuit
A logical circuit used in Frame Relay and X.25 WANs that provides communication between two network devices, and can be either permanent or switched.

virtual LAN (VLAN)
A LAN in which devices are logically configured to communicate as if they were attached to the same network, without regard to their physical locations.

voltage
The electromotive force that moves electrical current against resistance.

Glossary

waveform

The shape of a signal that has been graphed according to its amplitude and frequency during a specified period of time.

wide area network (WAN)

Two or more LANs or MANs that are interconnected using relatively slow-speed connections over telephone lines.

Windows 95

An operating system developed by Microsoft to run on PC-compatible computers. Released in 1995, it unveiled a more-friendly GUI, enhancements for networking, and a hardware auto-detect feature called Plug and Play.

Windows 98

A Microsoft operating system that retains the essential parts of Windows 95, but with new applications for Internet use and improved hardware recognition.

Windows NT Workstation 4

Microsoft's premiere desktop operating system, NT Workstation is essentially Windows NT Server. It has fewer network management applications and services than the server version, and each workstation is limited to 10 simultaneous connections and one remote access connection.

Windows XP Home Edition

Micrsoft's latest version of its Windows desktop operating system which has been trimmed down for home users. Many of the features of XP Professional have been removed from XP Home Edition.

Windows XP Professional Edition

Like XP Home Edition, XP Professional is Microsoft's latest operating system for the business environment. Many more networking features are available in the Professional Edition.

Wireless Access Point (WAP)

A network device that follows the IEEE 802.11a, 802.11b, or 802.11g protocols for providing network access to devices using radio frequencies.

wireless bridge

A device that connects LAN segments using wireless transmissions.

Wireless Equivalent Privacy

An encryption method used in wireless networks; typically called WEP.

WLAN

Wireless Local Area Network.

workgroup

The description of a logical group of users organized by job type.

workstation

Any computer that is capable of processing and storing user data for work-related tasks.

WPAN

Wireless Personal Area Network. WPAN devices include PCs, PDAs (Personal Digital Assistants), cell phones, pagers, and much more.

wrist strap

Also known as an *electrostatic wrist guard*, the wrist strap is worn around the wrist with a cable that connects to the metal case of the computer. Any static electricity from your body is then "grounded," preventing damage to the electronics equipment.

X.25

A wide area networking standard that defines how devices communicate within public data networks using data terminal equipment and data communications equipment.

Xerox Network System (XNS)

One of the first protocols used with Ethernet, XNS became the model used by many vendors for developing protocols.

zone

A group of network devices within an AppleTalk network.

Index

Note to the Reader: Throughout this index **boldfaced** page numbers indicate primary discussions of a topic. *Italicized* page numbers indicate illustrations.

Numbers

10Base2, **155–156**
10BaseT, **46–47**, *47*
568-A/UTP standard, **220**

A

abbreviations, listed alphabetically, 350–364
abstract syntax, defined, 22
AC (alternating current), **118–121**, *118–121*
access points (APs)
 basics of, **191**, *191*, 200
 defined, 4, 112, 182, 217
access rights (permissions), 57
ack, defined, 24
acronyms, listed alphabetically, 350–364
ad-hoc mode, defined, 112, *112*, 191
address gateways, defined, 199
Address Resolution Protocol (ARP), 185, **239–240**
addresses
 AppleTalk protocol suite and, **249**
 configuring Internet addresses, **89–93**, *90*, *92–93*
 Ethernet addresses, 262, *262*
 IP addressing, **241–242**, *241*
 IPX/SPX protocol suite and, 247
 MAC addresses (physical addresses), 31, *31*, 239
 network addresses, defined, 28
 private address space, defined, 91
 static addressing, defined, 72
agent software, defined, 301
ALOHAnet, 10
alternating current (AC), **118–121**, *118–121*
AM (amplitude modulation) radio signals, **137–138**, *137–138*
amp (ampere), defined, 136, *136*
analog, defined, 32
analog signals
 basics of, **137–138**, *137*
 vs. digital, **141–142**, *142*
ANSI (American National Standards Institute)
 ANSI WAN standards, **226–229**, *227*
 basics of, **211–212**
AppleShare, peer-to-peer networks and, 55
AppleTalk Address Resolution Protocol (AARP), 250
AppleTalk Data Stream Protocol (ADSP), **251**
AppleTalk Filing Protocol (AFP), 252
AppleTalk protocol suite, **249–252**
AppleTalk Transaction Protocol (ATP), **251**
application gateways, defined, 199
Application layer
 IPX/SPX protocol suite and, **248**
 as OSI model layer 7, **20–22**, *20*
 TCP/IP model and, **34–35**, **243–245**
 troubleshooting network application errors, **307–308**, *307*
application services, 21
architecture. *See also* network architectures
 selection of for LANs design, **267–268**
archive bits, defined, 288

ARCNET, defined, 10
ARP tables, 185
ARPAnet, **10**, 11
Asynchronous Transfer Mode (ATM)
 ATM switches, defined, 6
 basics of, **227–228**
asynchronous transmissions, **144**
AT&T, 8, 80, *80*
attenuation, defined, 276
authentication, defined, 87
autoloader, 287

B

backbones, defined, 11, 101
backing up, **286–289**, *289*
bandwidth
 basics of, **139–141**, *141*
 of various media, listed, 172, 175–176
baseband, defined, 155
baselining, **293**
batteries, 122
binary, defined, 32
BIOS (Basic Input/Output System), defined, 262
BISDN (Broadband ISDN), defined, 227
bit synchronization, defined, 144
bits
 archive bits, defined, 288
 vs. bytes, 140
 defined, 139
bit streams, defined, 139, *139*
blackouts, **127**
Bluetooth, defined, 219
BNC connectors, 100
bridges
 basics of, **192–193**, *192–193*, 200
 defined, 4, 5
broadband LANs standard, **216–217**
broadcast domains, defined, 183
brouters, **196**, *196*, 200
BSD UNIX, 79, *80*
bus topology, **100–101**, *100*
bytes, vs. bits, 140

C

cable
 cable Internet technology, **323–324**
 coaxial cable, **153–156**, *153*, *155*
 crossover cable, defined, 190
 LANs design and, **271–272**, **276–277**, 277
 single-mode cable, defined, 165
campus area networks (CANs), defined, 3
CCITT/ITU-T WAN standards, **223–226**
cell switching, defined, 320
central processing unit (CPU) usage, monitoring, 296
channels, B and D, 226
checksum, defined, 243
chips, defined, 171
cladding, defined, 164, *165*
client-server networks. *See also* network services and software
 architecture
  advantages and disadvantages of, 62
  basics of, 42, **59**, *59*
  common server types, **60**
  implementing, **63–66**, *64–66*
  LANs design and, 267
  security and, **62–63**, *63*
  selecting, 62
  terminals and hosts and, **61**, *61*
 configuration, **87–93**
  Internet addresses, **89–93**, *90*, *92–93*
  Windows XP, **87–89**, *88*
 described, 59, *59*
coaxial cable, **153–156**, *153*, *155*
collision detection, **214–215**, *215*
collision domains, defined, 183
commercial buildings cabling standards, **221–222**, *221–222*
Common Management Information Protocol (CMIP), **303**
Compatible Time-Sharing System (CTSS), 9
computers
 destination computers, defined, 2

mainframes, defined, 9, 61, *61*
source computers, defined, 2
connection-oriented transmissions, **26**, *27*
connectionless transmissions, **26–27**, *27*
connectors, **152**, **166–167**, *166*
continuity tests, defined, 276
copper media. *See under* network media
CPU usage, monitoring, 296
Crocker, Steve, 207
crossover cable, defined, 190
CSMA/CD (Carrier Sense Multiple Access/Collision Detection) access method, **214–215**, *214*

D

daisy chains, defined, 269
data
    backing up, **286–289**, *289*
    binary data, defined, 32
    data compression, 22–23
    data over cable technology, **323–324**
    defined, 18
    OSI model and, **18**, *18*, 22
data centers, **128**
Data Link layer (layer 2)
    basics of, **30–31**, *30–31*
    protocols, **238–239**, **250**, 308, *308*
database services, 21
Datagram Delivery Protocol (DDP), **251**
datagrams, defined, 28
DC (direct current), **121–122**, *121*
deencapsulation, defined, 18
default gateways, defined, 89
demarcation point, defined, 264
demultiplexing, defined, 317
devices, **181–203**
    access points, **191**, *191*
    bridges, **192–193**, *192–193*
    brouters, **196**, *196*
    comparing, **200–201**
    defined, 2, 16
    extending networks and, **182**, *182*
    gateways, **199**

hubs, **189–190**, *189*
LANs
    connecting, **277–278**
    selecting, **273–275**
network interface cards (NICs), **185–187**, *186*
network segments, **183–184**, *184*
repeaters, **188**, *188*
review questions, 202–203
routers, **197–198**, *198*
switches, **194–195**, *194–195*
DHCP (Dynamic Host Configuration Protocol), 72, **245**
dialog control, **25**, *25*
dialog separation, 25
dielectric, 153, *153*, 157, *157*
digital, defined, 32
Digital Linear Tape (DLT), 286
digital signals
    analog vs. digital, **141–142**, *142*
    basics of, **138–141**, *139*, *141*
Digital Subscriber Line (DSL), **321–323**, *322–323*, 324
direct current (DC), **121–122**, *121*
direct-sequence (DS) modulation, defined, 171
disk mirroring, 292
disk striping with parity, 292
Distributed Queue Dual Bus (DQDB) access method standard, **216**
distribution layer, 269
documentation
    LAN design and, **263–265**
    network management and, **296–298**, *297*, **305–306**
DoD model. *See* TCP/IP
domain name registry, defined, 327
Domain Name System (DNS), 72, **91**, **244–245**
domains, defined, 91
downtime, defined, 283
DQDB (Distributed Queue Dual Bus) access method standard, **216**
drivers, defined, 185
DSO, 319

dummy terminals, defined, 61
Dynamic Host Configuration Protocol (DHCP), 72, **245**

E

e-mail, Application layer and, 21
earth grounds, **119–120**, *120*
electricity, **117–131**
    data centers and, **128**
    power problems and solutions, **123–127**
    review questions, 129–131
    types of current, **118–122**
        alternating (AC), **118–121**, *118–121*
        direct (DC), **121–122**, *121*
electromagnetic interference (EMI), 134–135, 156
Electronics Industry Alliance (EIA)
    about, **210**
    EIA/TIA cabling standards, **220–222**, *221–222*
encapsulation, 18, *18*, 29
encryption, 23–24
enterprise networks, **65–66**, *66*
equipment, assessing for LANs design, **260–263**, *262*
Ethernet, defined, 10, 101

F

Fast Ethernet, **215**
fault tolerance, 76, 106, 285
faults, defined, 296
FDDI (Fiber Distributed Data Interface), 227
Federal Communications Commission (FCC), 8
feedback loops, defined, 259
Fiber Distributed Data Interface (FDDI), **226–227**
fiber-optic media, *165–167*
fiber optics
    connections, 3

fiber-optic media, **164–168**, *165–167*
    standards for, **217**
file services, Application layer and, 21
file sharing. *See* folders
File Transfer Protocol (FTP), 20, 72, **244**
firewalls, defined, 278
flow control, **27–28**, *27*
FM (frequency modulation) radio signals, **137–138**, *138*
folders
    sharing in Macintosh, **55–58**, *55–58*
    sharing in Windows, **51–-55**, *51–55*
Frame Relay, **224–225**, **324–325**
frequency
    defined, 136, *136*
    frequency hopping, defined, 171
    wireless ranges, 172
frequency modulation (FM) radio signals, **137–138**, *138*
full-duplex transmissions (FDX), **146–147**, *147*

G

gateways, **199**, 200
Get Nearest Server (GNS), 247
Gigabit Ethernet, **215**
gigabit Points of Presence (gigaPOPs), defined, 11
glossary of terms, 366–383

H

half-duplex transmissions (HDX), **146**, *146*
hardware
    identifying for LANs design, **262–263**, *262*
    server hardware, **76**
hertz (Hz), defined, 136
hexadecimal addresses, 31, *31*
high-powered radio transmissions, **170–171**
horizontal cross-connects, defined, 162
Host-to-Host layer (TCP/IP model), 35
hosts and terminals, **61**, *61*

387

hot wires, defined, 120, *120*
hubs
    basics of, 4, *4*, 33, *33*, **189–190**, *189*, 200
    star topology and, 104, *104*, 105, *105*, 108, *108*
hybrid networks, **67–68**, *68*
hybrid topologies, **108**, *108*, 267
HyperText Transfer Protocol (HTTP), 20, **244**

I

IBM data connectors, 159, *159*
IEEE (Institute of Electrical and Electronics Engineers)
    about, **210**
    IEEE 802: LANs and MANs standards, **213–219**, *214–215*
    Web site for, 218
impedance, and coaxial cables, 154
infrared transmissions, **174**
infrastructure mode, 112
infrastructure mode, defined, *112*, 191
Integrated Service Digital Network (ISDN), **325–326**
integrated services (IS) standards, **217**
interference, 134–135
intermediate distribution facilities (IDFs), **270–271**, *270*
intermediate systems, 29
International Organization for Standardization (ISO), 16, 209
International Telecommunications Union (ITU), **212**
the Internet
    basics of, **11**
    Internet 2
        defined, 11, 228
        Web site for, 11, 229
    Internet access. *See* wide area networks (WANs) and Internet access
Internet Control Message Protocol (ICMP), **240**

Internet layer (TCP/IP) model, **35**, **239–242**, *241–242*
Internet Packet eXchange, **247–248**
Internet Protocol (IP)
    basics of, **239**
    Version 6, **242**
Internet Service Providers (ISPs), connecting to, **317**
internetworks, defined, 2
interoperability, TCP/IP suite and, **238**
IP addressing, **241–242**, *241*
IP subnets, **242**, *242*
IPX/SPX protocol suite, **246–248**, *246*
IS (integrated services) standards, **217**
ISA (Industry Standard Architecture), network connections and, 47

J

joules, defined, 125

K

keep-alive messages, 24

L

LANs. *See* local area networks (LANs)
latency, defined, 283
layers, OSI model. *See also under* OSI model
    network troubleshooting and, **306–310**, *307–309*
leased lines, defined, 319
LEDs (light-emitting diodes), defined, 164
line conditioners, defined, 127
Linux, *80*, **81–83**
local area networks (LANs)
    basics of, **3–5**, *4*
    IEEE 802: LANs and MANs standards, **213–219**, *214–215*
    network segments, 183, **184**, *184*
    virtual LANs, **194–195**, *195*
    wireless LAN topologies, **112**, *112*

local area networks (LANs) design, **257–279**
  creating, **267–278**
    architecture selection, **267–268**
    cable, testing and certifying, **276–277**
    device selection, **273–275**
    LANs devices, connecting, **277–278**, *277–278*
    media selection and installation, **275–276**
    topology selection, **269–272**, *269–270*
  needs assessment, **260–266**
    equipment inventory, **260–263**, *262*
    facility assessment and documentation, **264–265**, *264*
    user needs, **266**
  preparation for, **258–259**, *258–259*
  review questions, 279
local exchange carrier (LEC), defined, 316
local loops, defined, 316
LocalTalk protocol, 250
Logical Link Control (LLC)
  802.2 standards, **213**
  LLC sublayer, 30, *30*
logical topologies. *See under* topologies

M

MAC broadcasts, defined, 240
Macintosh
  configuring peer-to-peer networks for, **55–58**, *55–58*
  OS X
    identifying system information in, **262–263**, *262*
    server, **85–86**
main distribution facility (MDF), 269–270
mainframes, defined, 9, 61, *61*
management information base (MIB), defined, 300
management of networks. *See* network management
MANs. *See* metropolitan area networks (MANs)
MAUs (Multistation Access Units), 108, *108*, 110, *110*, 190

media. *See* network media
Media Access Control (MAC) sublayer, 30–31, *30*
memory in cassette (MIC), defined, 286
mesh topology, **106–107**, *106–107*
metal oxide varistor (MOV), defined, 125
metropolitan area networks (MANs)
  basics of, **5–6**
  IEEE 802: LANs and MANs standards, **213–219**, *214–215*
  security standards, **217**
metropolitan exchange, defined, 316
MIC (memory in cassette), defined, 286
microwave transmissions, **172–174**, *173*
modems, 7, 142, *142*
motherboards, defined, 185
multi-server networks, **64–65**, *65*
multichannel trunk, defined, 317
multimode cable, defined, 165
multiplex, defined, 317
multiplexing, defined, 217
multiprocessing, and network services, **75**, *75*, 76
multitasking, and NOSs, **74**, *74*

N

Name Binding Protocol (NBP), **251**
narrow-band radio transmissions, **170**
NDS (Novell Directory Services), 83–84
near-end cross talk, defined, 276
needs assessment, defined, 260
NetBEUI (NetBIOS Extended User Interface), defined, 89
NetWare, **83–84**
NetWare Core Protocol (NCP), **248**
network architectures
  basics of, 42
  client-server networks. *See* client-server networks, architecture
  hybrid networks, 42, **67–68**, *68*
  peer-to-peer networks. *See* peer-to-peer networks, architecture
  review questions, 69

389

network interface cards (NICs)
  defined, 4, 5
  devices and, **185–187**, *186*
  MAC addresses and, 31
Network Interface layer, **35**
Network layer
  IPX/SPX protocol suite and, **247–248**
  OSI model (layer 3), **28–29**, *28*
  TCP/IP model, **35**, **239–242**, *241–242*
  troubleshooting, 308, *308*
network management, **281–311**
  first steps in, **285–292**
    backing up, **286–289**, *289*
    basics of, **285**
    redundancy, **291–292**
    uninterrupted power, **290**
  importance of, **282–284**, *282–283*
  network management systems (NMS), **299–304**
    implementing, **303–304**
    SNMP and, **300–303**, *301*
    TCP/IP networks, **299–300**, *300*
  performance monitoring, **293–298**
    baselining, **293**
    documenting performance, **296–298**, *297*
    network performance and, **294–295**, *295*
    server performance and, **296**
  review questions, 311
  troubleshooting, **305–310**
    documentation for, **305–306**
    layered approach, **306–310**, *307–309*
network media, **151–179**
  comparing, **175–176**
  connectors and, **152**
  copper media, **153–163**
    coaxial cable, **153–156**, *153*, *155*
    shielded twisted-pair cable, **157–159**, *157*, *159*
    unshielded twisted pair, **160–163**, *160*, *162*

defined, 30, *30*, 98, 152
fiber-optic media, **164–168**, *165–167*
review questions, 177–179
selection of for LANs design, **275–276**
troubleshooting, **309–310**, *309*
wireless networking options, **169–174**, *169*
  infrared transmissions, **174**
  microwave transmissions, **172–174**, *173*
  radio transmissions, **170–172**
network operating systems (NOSs). *See under* network services and software
network operations center (NOC), defined, 304
network-printing services, 21
network protocols, **233–255**
  AppleTalk protocol suite, **249–252**
  defined, 2
  importance of, **234–235**
  IPX/SPX protocol suite, **246–248**, *246*
  protocol suites basics, **236**
  review questions, **253–255**
  TCP/IP suite, **237–245**, *237*
    Application layer, **243–245**
    Data Link layer, **238–239**
    features of, **238**
    Internet (Network) layer, **239–242**, *241–242*
    Transport layer, **243**
    troubleshooting, **308–309**, *308*
network services and software, **71–95**
  client network configuration, **87–93**
    Internet addresses, **89–93**, *90*, *92–93*
    Windows XP, **87–89**, *88*
  network operating systems (NOSs)
    basics of, **74–75**, *74–75*
    Linux, **81–83**
    Mac OS X Server, **85–86**
    Novell NetWare, **83–84**
    UNIX, **79–80**, *80*
    Windows 2000 Server, **77–79**, *78*
  review questions, 94–95
  servers, selecting, **74–77**, *74–75*

networks
    basics of, **1–13**
        fundamentals, **2**
        the Internet, **11**
        origins of, **9–10**
        public telephone system and, **8**
        review questions, 12–13
        types of, **3–7**, *4*, *6*
    client-server networks. *See* client-server networks
    defined, 1
    network segments, **183–184**, *184*
NOC (network operations center), defined, 304
nodes, defined, 2
Novell Directory Services (NDS), 83–84
Novell NetWare, **83–84**
NSFNET, 11

O

ohms, defined, 154
operating systems. *See under* network services and software
orthogonal frequency division multiplexing (OFDM), defined, 171
OS X
    identifying system information in, **262–263**, *262*
    server, **85–86**
OSI (Open Systems Interconnection) model, **16–39**
    basics of, **16–17**, *16–17*
    data flow and, **18**, *18*
    OSI layer basics, **19–33**, *19*
        Application layer (layer 7), **20–22**, *20*
        Data Link layer (layer 2), **30–31**, *30–31*
        Network layer (layer 3), **28–29**, *28*
        Physical layer (layer 1), **32–33**, *32–33*
        Presentation layer (layer 6), **22–24**, *23*
        Session layer (layer 5), **24–25**, *25*
        Transport layer (layer 4), **26–28**, *27*
    review questions, 37–39
    vs. TCP/IP model, **34–36**

P

packets
    defined, 18, 22
    packet switching, **320**
passwords, and share-level security, 45
patch cable, defined, 277
PCI (Peripheral Component Interconnect), 47
PDAs (Personal Digital Assistants), defined, 219
PDN (public data network), defined, 223
peer-to-peer networks
    architecture
        basics of, 42
        implementing, **46–47**, *47*
        LANs design and, 267
        Macintosh and, **55–58**, *55–58*
        security on, **45**, *45*
        selecting, **44**
        Windows and, **48–55**, *49–54*
    peer-to-peer mode, 112, *112*
Performance Monitor, configuring, **297**, *297*
performance monitoring. *See under* network management
Peripheral Component Interconnect (PCI), 47
permissions, in Macintosh networks, 57
Physical layer
    OSI model, **32–33**, *32–33*
    TCP/IP model, **36**
    troubleshooting, 309–310, *309*
physical topologies. *See under* topologies
ping, 92–93, 309
plenum, defined, 153, *165*, 275
point of entry (POE), defined, 264
Point-to-Point Protocol (PPP), **321**
polling, defined, 302
ports, defined, 189, 190

power conditioners, defined, 127
premise, defined, 316
Presentation layer (layer 6), **22–24**, *23*
printing. *See* network-printing services
Process layer (TCP/IP model), **34–35**
promiscuous mode, defined, 302
protocol gateways, defined, 199
protocols. *See* network protocols
public data network (PDN), defined, 223
public switched telephone network (PSTN), 8, 223
public telephone system
  networking basics and, **8**
  WANs and, 6
punch block, defined, 275

R

radio
  radio frequency interference (RFI), 134–135, 156
  radio signals, **137–138**, *138*
  radio transmissions, **170–172**
redundancy, 103, 285, **291–292**
Redundant Array of Inexpensive Disks (RAID), 76, 292
remote network monitoring, **302–303**
repeaters, 33, *33*, **188**, *188*, 200
Request for Comments (RFC), defined, 206
Reverse Address Resolution Protocol (RARP), **240**
review questions and answers
  devices, 202–203, 342–343
  electricity, 129–131, 339
  local area networks (LANs) design, 279, 345–346
  network architectures, 69, 334–335
  network management, 311, 346–347
  network media, 177–179, 340–341
  network protocols, 253–255, 344–345
  network services and software, 94–95, 336–337
  networking basics, 12–13, 332
  OSI model, 37–39, 333–334
  signaling, 148–149, 340
  standards, 230–231, 343–344
  topologies, 113–115, 337–338
  wide area networks (WANs) and Internet access, 329–330, 347–348
RFI (radio frequency interference), 134–135, 156
ring topology, **101–104**, *102–103*
routers
  basics of, **197–198**, *198*, 200
  configuring DSL routers, **322–323**, *322–323*
  defined, 4, 5
  selection of for LANs design, **274**
routing, 29, 197. *See also* routers

S

sags, **126–127**
satellites
  satellite microwave, **174**
  WANs and, 6
SC connectors, **167**, *167*
SCSI (Small Computer System Interface), defined, 286
security
  client-server networks and, **62–63**, *63*
  LAN/MAN standards, 217
  LANs design and, 266, 271
  peer-to-peer networks and, **45**, *45*
  share-level, 44, 45
Sequence Packet eXchange, **248**
serial ports, defined, 46
servers
  common server types, **60**
  selecting, **74–76**, *74–75*
  server-based networks. *See* client-server networks, architecture
Service Advertisement Protocol (SAP), **248**
services
  Application layer services, 21

integrated services (IS) standards, **217**
on servers, defined, 59, 71. *See also* network services and software
types of, **72–73**
Session layer (layer 5), **24–25**, *25*
sessions, defined, 24
shielded twisted-pair (STP) cable, **157–159**, *157*, *159*
signaling, **133–149**
  basics of, **134–135**
  computers and, **137–143**
    analog, **137–138**, *137*
    analog vs. digital, **141–142**, *142*
    digital, **138–141**, *139*, *141*
  measuring signals, **136**, *136*
  review questions, 148–149
  transmission, **144–147**, *145–147*
Simple Mail Transfer Protocol (SMTP), 20, **244**
Simple Network Management Protocol (SNMP), **245**
simplex transmissions, **145**, *145*
sine waves, 118, *118*, 136, *136*, 137
single-mode cable, defined, 165
single-server networks, **64**, *64*
SMA connectors, **166**, *166*
Small Computer System Interface (SCSI), defined, 286
SNMP, **300–303**, *301*
software
  network operating systems (NOSs). *See under* network services and software
  network services available with, **72–73**
SONET (Synchronous Optical Network), **228–229**
Spanning Tree Protocol (STP), 213
spikes, 125, *125*
spread-spectrum radio transmissions, **171–172**, *171*
square waves, 136, *136*, 138, *138*
ST connectors, **166**, *166*
stacks, defined, 234

standards, **205–231**
  basics of, **206–207**
  EIA/TIA cabling standards, **220–222**, *221–222*
  IEEE 802: LANs and MANs standards, **213–219**, *214–215*
  review questions, 230–231
  standards organizations, **208–212**
  WAN connection standards, **223–229**
    ANSI WAN standards, **226–229**, *227*
    CCITT/ITU-T WAN standards, **223–226**
star topology, **104–106**, *104–105*
static electricity, **123**
STP. *See also* shielded twisted-pair (STP) cable
  Spanning Tree Protocol (STP), 213
subnets, **242**, *242*
Subnetwork Access Protocol (SNAP), 213
SuperDLT, 287
surge suppressors, **125–126**, *126*
surges, **124–126**, *124*, *126*
swap file errors, 296
switches
  basics of, 4, *4*, **194–195**, *194–195*, 200
  selection of for LANs design, **273**
syn, defined, 24
Synchronous Optical Network (SONET), **228–229**
synchronous transmissions, **144–145**
syntax, abstract and transfer, 22

T

T-carriers, defined, 319
tape drives
  tape backup drives, 76
  types, listed, 286–287
TCP/IP model
  configuring with Windows XP, **89–91**, *90*, **92–93**, *92–93*
  management model for, **299–300**, *300*

vs. OSI model, **34–36**
TCP/IP suite. *See under* network protocols
Telecommunications Industry Association. *See* TIA (Telecommunications Industry Association)
telecommunications services, **319–320**
telephones. *See* public telephone system
Telnet, **245**
terminals and hosts, **61**, *61*
terminators, 100
terrestrial microwave, **173**, *173*
Thicknet, **155**, *155*
Thinnet, **155–156**
TIA (Telecommunications Industry Association)
    about, **211**
    EIA/TIA cabling standards, **220–222**, *221–222*
Time Domain Reflectometry (TDR), defined, 272
Token Ring, 10, 102, **216**
tokens, defined, 216
topologies, **97–115**
    logical, **109–111**, *109*
        logical bus, **109–110**, *109*
        logical ring, **110–111**, *110*
    physical, **99–108**, *99*
        bus, **100–101**, *100*
        hybrid, **108**, *108*
        mesh, **106–107**, *106–107*
        ring, **101–104**, *102–103*
        star, **104–106**, *104–105*
    physical vs. logical, **98**
    selection of for LANs design, **269–272**, *269–270*
    wireless LAN topologies, **112**, *112*
Torvalds, Linus, 81
Trace Routes, **93**, *93*
traceroute, and troubleshooting, 309
transfer syntax, defined, 22
transmission
    basics of, **144–147**, *145–147*
    infrared, **174**

microwave, **172–174**, *173*
radio, **170–172**
rates for SONET level, listed, 229
Transmission Control Protocol (TCP), **243**
transponders, defined, 174
Transport layer
    IPX/SPX protocol suite and, **248**
    OSI model (layer 4), **26–28**, *27*
    TCP/IP model, **35**, **243**
    troubleshooting, 308, *308*
Trivial File Transfer Protocol (TFTP), **244**
troubleshooting, **305–310**
    documentation for, **305–306**
    layered approach, **306–310**, *307–309*
TV, access method standard, **218**

U

Ultrium (LTO), 286
uninterruptible power supply (UPS), 76, 126, **290**
UNIX. *See also* Linux
    as network operating system, **79–80**, *80*
unshielded twisted pair (UTP), **160–163**, *160, 162*
uplink ports, defined, 190
User Datagram Protocol (UDP), **243**
users
    assessing needs for LANs design, **266**
    Macintosh and definition of, 55

V

Vail, Theodore, 8
vampire tap, defined, 155
vBNSs (very high performance Backbone Network Services), defined, 11
virtual circuits, defined, 224
virtual private network (VPN) services, 73

W

waveforms, 136, *136*. *See also* sine waves

Web servers, 60
Web sites for information
  downloading Linux, 82
  IEEE, 218
  Internet 2, 11, 229
  RFC Archive, 206
wide area networks (WANs)
  basics of, **6–7**, *6*, **314–315**
  connecting to, **278**, *278*
  WAN connection standards, **223–229**
    ANSI WAN standards, **226–229**, *227*
    CCITT/ITU-T WAN standards, **223–226**
wide area networks (WANs) and Internet access, 313–330
  connecting to the Internet, **316–317**, *316–317*
  planning Internet access, **318–328**
    avoiding potholes, **326**
    basics of, **318**
    data over cable, **323–324**
    DSL, **321–323**, *322–323*
    DSL vs. cable Internet, 324
    Frame Relay, **324–325**
    Integrated Service Digital Network (ISDN), **325–326**
    point-to-point protocol (PPP), **321**
    the rollout, **326–328**
    telecommunications services, **319–320**
  review questions, 329–330
  WANs basics, **314–315**

Windows 2000
  connecting to Windows 2000 networks, **87–88**
  Windows 2000 Server, as network operating system, **77–79**, *78*
Windows, configuring peer-to-peer networks for, **48–55**, *49–54*
Windows XP
  client network configuration and, **87–89**, *88*
  identifying system information in, **262**
Wireless Access Points (WAPs), LANs design and, 265, **274**
wireless bridges, 182, **193**, *193*, 200
Wireless Equivalent Privacy (WEP), defined, 191
wireless networking options. *See* network media
Wireless Personal Area Network (WPAN) standards, **218–219**
workgroups, defined, 43, *43*
workstations
  defined, 42
  single-server networks and, 64, *64*
wrist straps, 47

X

X.25, **224**
Xerox Network System (XNS), **246**

Z

Zone Information Protocol (ZIP), 252
zones, defined, 250

The Complete CCNA Study Solution

Sybex has CCNA certification covered. From Study Guides to Sybex Virtual Lab™ and Virtual Trainer™ software, we have the training materials you need to approach the exams with confidence.

CCNA: Cisco Certified Network Associate Study Guide, 3rd Edition
ISBN: 0-7821-4167-6 • $49.99

CCNA: Cisco Certified Network Associate Study Guide, Deluxe Edition, 2nd Edition
ISBN: 0-7821-4169-2 • $89.99

CCNA : Cisco Certified Network Associate Exam Notes
ISBN: 0-7821-4168-4 • $29.99

CCNA Virtual Training Certification Kit
ISBN: 0-7821-3033-X • $139.99

CCNA Virtual Lab, Gold Edition
ISBN: 0-7821-3018-6 • $149.99

CCNA Certification Kit 2nd Edition
ISBN: 0-7821-4170-6
$159.99

www.sybex.com